Synthesis Lectures on Mathematics & Statistics

Series Editor

Steven G. Krantz, Department of Mathematics, Washington University, Saint Louis, MO, USA

This series includes titles in applied mathematics and statistics for cross-disciplinary STEM professionals, educators, researchers, and students. The series focuses on new and traditional techniques to develop mathematical knowledge and skills, an understanding of core mathematical reasoning, and the ability to utilize data in specific applications.

Sujaul Chowdhury · Abdullah Al Sakib

Numerical Exploration of Fourier Transform and Fourier Series

The Power Spectrum of Driven Damped Oscillators

 Springer

Sujaul Chowdhury
Department of Physics
Shahjalal University of Science
and Technology
Sylhet, Bangladesh

Abdullah Al Sakib
Department of Physics
Shahjalal University of Science
and Technology
Sylhet, Bangladesh

ISSN 1938-1743   ISSN 1938-1751 (electronic)
Synthesis Lectures on Mathematics & Statistics
ISBN 978-3-031-34663-7   ISBN 978-3-031-34664-4 (eBook)
https://doi.org/10.1007/978-3-031-34664-4

This Springer imprint is published by the registered company Springer Nature Switzerland AG
The registered company address is: Gewerbestrasse 11, 6330 Cham, Switzerland

Preface

This book contains practical demonstrations of numerically obtaining the Fourier transform of given numerical data. In particular, we demonstrate how to obtain the frequencies that are present in data numerically, using what is called discrete Fourier transform. We have used programs written in Mathematica in this regard.

To obtain the Fourier series, we first need to know the frequencies that we need to use. Here lies a need for the Fourier transform. We find that if we have the Fourier series for a function $y(t)$, we can plot the Fourier series in an extended interval of time t. We find that a function need not be periodic to be expressed analytically as a Fourier series. But after expressing the function as a Fourier series, we can plot it in an extended interval of time and get repeated or periodic plots of the original non-periodic function.

This book also contains numerical solutions of differential equations of driven damped oscillators using the 4th-order Runge-Kutta method using programs written in Mathematica. Data of the numerical solution are compared with analytical solutions and are fed to a discrete Fourier transform program to obtain frequency content of the oscillator using programs written in Mathematica.

The behavior of mechanical systems such as driven damped oscillators can be depicted by plotting velocity versus displacement, rather than displaying displacement as a function of time. This velocity versus displacement coordinate system is known as phase space. The trajectory in phase space provides another perspective of a system, and often it is more valuable than the displacement versus time plot. We have depicted the motion of a simple harmonic oscillator, a damped harmonic oscillator and a driven damped oscillator in phase space.

This book contains the first and first-hand practical demonstrations of obtaining discrete Fourier transform of data numerically using Mathematica. We have explored the use of discrete Fourier transform using Mathematica. Besides graduate students for the course titled Computational Physics, this book will prove useful to all of Physical Science and Engineering who often need to know the frequencies that are present in numerical data.

Sylhet, Bangladesh   Sujaul Chowdhury
2023   Abdullah Al Sakib
                                                                         www.sust.edu

Contents

About the Authors

Sujaul Chowdhury Ph.D., is a Professor in Department of Physics at Shahjalal University of Science and Technology. He also received his B.Sc. and M.Sc. in Physics from Shahjalal University of Science and Technology, before earning his Ph.D. at The University of Glasgow. After completing his Ph.D., Dr. Chowdhury was a Humboldt Research Fellow for one year at Max Planck Institute. He is the author of many books, including *Monte Carlo Methods: A Hands-On Computational Introduction Utilizing Excel* and *Monte Carlo Methods Utilizing Mathematica®: Applications in Inverse Transform and Acceptance-Rejection Sampling*, which are also in the Synthesis Lectures on Mathematics & Statistics series published by Springer Nature. His research interests include nanoelectronics, magnetotransport in semiconductor nanostructures and nanostructure physics.

Abdullah Al Sakib is an M.S. student in the Department of Physics at Shahjalal University of Science and Technology.

Exploring Fourier Transform and Fourier Series Approximation Numerically

Abstract

This chapter contains practical demonstrations on numerically obtaining Fourier transform of given numerical data. In particular, we demonstrate how to obtain the frequencies that are present in the data numerically using what is called discrete Fourier transform. We also demonstrate how to numerically obtain Fourier series approximation to any function. Programs were written in Mathematica in this regard.

1.1 Frequency Content in Oscillatory Motion

Suppose, we have a given set of numerical data, for an oscillatory motion. For example, suppose, we have a set of values of displacement y of a particle for a given set of values of time t. We can numerically get the frequencies that are present in the data by obtaining Fourier transform. This is outlined in the following.

Fourier transform pair is given by

$$y(t) = \frac{1}{\sqrt{2\pi}} \int_{-\infty}^{+\infty} A e^{i\omega t} \, d\omega \tag{1.1}$$

and

$$A = \frac{1}{\sqrt{2\pi}} \int_{-\infty}^{+\infty} y(t) \, e^{-i\omega t} \, dt \tag{1.2}$$

© The Author(s), under exclusive license to Springer Nature Switzerland AG 2024
S. Chowdhury and A. Al Sakib, *Numerical Exploration of Fourier Transform and Fourier Series*, Synthesis Lectures on Mathematics & Statistics,
https://doi.org/10.1007/978-3-031-34664-4_1

Equation (1.2) tells us about frequency content in Eq. (1.1). In other words, Eq. (1.2) tells us which frequencies are present in the data for y and relative importance or dominance of these frequencies.

To find A numerically, we replace Eq. (1.2) by

$$A = \frac{1}{\sqrt{2\pi}} \int_0^T y(t)\, e^{-i\omega t}\, dt \tag{1.3}$$

as an approximation. Here 0 to T is a time interval in which we have a set of numerical data for $y(t)$. T need not be true period in the oscillatory motion. We take $y(t) = y(t + T)$.

We know from *trapezoidal rule* for numerical integration that

$$\int_0^T v\, dt = \frac{h}{2}\big[v_0 + 2(v_1 + v_2 + v_3 + \cdots + v_{n-1}) + v_n\big] \tag{1.4}$$

where v_i's are equally spaced values of v, in the interval 0 to T; h is the spacing. If $v_0 = v_n$, we get

$$\int_0^T v\, dt = h(v_1 + v_2 + v_3 + \cdots + v_n) \tag{1.5}$$

As such, Eq. (1.3) can now be written as

$$A(n) = h \frac{1}{\sqrt{2\pi}} \sum_{k=1}^N y_k\, e^{-i\omega_{n_1} t_k} \tag{1.6}$$

where y_k is value of y for $t = t_k = k\,h$, $\omega_{n_1} = n_1\, \omega_1 = n_1\,(2\pi/T)$ with $T = N\,h$. $n_1 = 0, 1, 2, 3, \ldots, N$. $\omega_0 = 0$ for $n_1 = 0$ corresponds to the zero frequency or dc component of the signal $y(t)$ that does not oscillate. $n = n_1/T$ Eq. (1.6) gives

$$A(n) = \frac{h}{\sqrt{2\pi}} \sum_{k=1}^N y_k\, e^{-i\left(n_1 \frac{2\pi}{Nh}\right)k\,h}$$

or,

$$A(n) = \frac{h}{\sqrt{2\pi}} \sum_{k=1}^N y_k\, e^{-i\left(\frac{2\pi}{N}\right)k n_1} \tag{1.7}$$

or,

$$A(n) = \frac{h}{\sqrt{2\pi}} \sum_{k=1}^{N} y_k \, Z^{kn_1} \qquad (1.8)$$

where

$$Z = e^{-i\left(\frac{2\pi}{N}\right)} \qquad (1.9)$$

Equations (1.1)–(1.9) have been adapted from Ref. [1].

If we calculate A using Eq. (1.7), we find non-zero frequency content for both $A(n)$ and $A(n_{max}-n)$. $A(n)$ is the physical result while that for $A(n_{max}-n)$ is unphysical, though mathematically valid as shown below. $n_{max} = N/T$ Eq. (1.7) gives

$$A(n_{max} - n) = \frac{h}{\sqrt{2\pi}} \sum_{k=1}^{N} y_k \, e^{-i\left(\frac{2\pi}{N}\right)k(N-n_1)}$$

$$= \frac{h}{\sqrt{2\pi}} \sum_{k=1}^{N} y_k \, e^{-i\,2\pi\,k} \, e^{i\left(\frac{2\pi}{N}\right)n_1 k}$$

$$= \frac{h}{\sqrt{2\pi}} \sum_{k=1}^{N} y_k \, e^{i\left(\frac{2\pi}{N}\right)n_1 k} \quad \text{because } e^{-i\,2\pi\,k} = 1 \text{ because k is integer}$$

$$= \frac{h}{\sqrt{2\pi}} \sum_{k=1}^{N} y_k^* \left(Z^*\right)^{n_1 k} \quad \text{because } y_k \text{ is real}$$

Thus

$$A(n_{max} - n) = A^*(n) \qquad (1.10)$$

leading to

$$|A(n_{max} - n)|^2 = |A(n)|^2 \qquad (1.11)$$

In the following sections, we have shown examples on practically calculating $A(n)$ for different data sets $y(t)$, using Eq. (1.8). We have demonstrated that frequencies that are present in the data can be obtained quantitatively and exactly. We can also get relative strengths or dominance of different frequencies in the data.

1.2   Discrete Fourier Transform: Example I

In this section, we take up the function $y = \cos(4\pi\,t) = \cos(2\pi\,2\,t)$. Using Program number 1.1, we first turn the analytic function $y(t)$ into numerical data denoted by $y[i]$ for a discrete set of values of time in the interval 0–10 s. These discrete data are then fed to the rest of the program that calculates (the complex numbers) $A[n]$'s, and absolute values of them which constitute what is called power spectrum (Table 1.1 and Fig. 1.1).

Table 1.1 Discrete Fourier transform for the numerical data for $y = \cos(2\pi \, 2 \, t)$

n	A[n]	Abs [A[n]]
0.0	4.69933*10^-15	4.69933*10^-15
0.1	6.55514*10^-16+1.13732*10^-15 I	1.31271*10^-15
0.2	-1.24016*10^-16+2.82164*10^-15 I	2.82437*10^-15
0.3	-5.00494*10^-16+1.05912*10^-15 I	1.17142*10^-15
0.4	-9.25692*10^-16+1.063*10^-16 I	9.31775*10^-16
0.5	-6.64372*10^-16-4.99387*10^-16 I	8.3113*10^-16
0.6	6.46656*10^-16-1.063*10^-16 I	6.55334*10^-16
0.7	-4.20769*10^-16+3.34401*10^-16 I	5.37467*10^-16
0.8	2.52461*10^-16+8.96903*10^-17 I	2.6792*10^-16
0.9	-9.83271*10^-16-1.78716*10^-15 I	2.0398*10^-15
1.0	2.97639*10^-15-3.33293*10^-16 I	2.99499*10^-15
1.1	-4.03053*10^-16+2.03741*10^-15 I	2.07689*10^-15
1.2	-5.44785*10^-16+2.36959*10^-15 I	2.43141*10^-15
1.3	2.12599*10^-16+2.74607*10^-15 I	2.75429*10^-15
1.4	1.63879*10^-16+1.94329*10^-15 I	1.95019*10^-15
1.5	5.1821*10^-16+1.90675*10^-15 I	1.97591*10^-15
1.6	2.23672*10^-15+2.55562*10^-15 I	3.39619*10^-15
1.7	4.53545*10^-15-1.01428*10^-15 I	4.64748*10^-15
1.8	5.30612*10^-15+2.09499*10^-15 I	5.70473*10^-15
1.9	2.94981*10^-15+4.53545*10^-15 I	5.41033*10^-15
2.0	**1.99471+1.42796*10^-14 I**	**1.99471**
2.1	-3.75149*10^-15-1.18258*10^-15 I	3.93347*10^-15
2.2	-3.84007*10^-15+9.21263*10^-16 I	3.94904*10^-15
2.3	-3.13584*10^-15-3.47688*10^-15 I	4.68211*10^-15
2.4	1.74508*10^-15+5.89077*10^-16 I	1.84183*10^-15
2.5	-5.37699*10^-15+1.13386*10^-15 I	5.49524*10^-15
2.6	1.98426*10^-15-2.57776*10^-15 I	3.25302*10^-15
2.7	-1.75394*10^-15+8.99117*10^-16 I	1.97097*10^-15
2.8	-2.44489*10^-15+1.24681*10^-15 I	2.74445*10^-15
2.9	-8.63684*10^-16+9.25692*10^-16 I	1.26604*10^-15

(continued)

Table 1.1 (continued)

n	A[n]	Abs [A[n]]
3.0	-6.01921*10^-15+2.34745*10^-16 I	6.02379*10^-15
3.1	1.85138*10^-15-1.61885*10^-15 I	2.45933*10^-15
3.2	-5.53644*10^-16-1.88239*10^-16 I	5.84769*10^-16
3.3	-1.13829*10^-15-7.92818*10^-16 I	1.38718*10^-15
3.4	-1.05414*10^-15+1.04749*10^-15 I	1.48609*10^-15
3.5	-1.27117*10^-15+1.45276*10^-15 I	1.93038*10^-15
3.6	2.05513*10^-15-1.17151*10^-15 I	2.36558*10^-15
3.7	-2.02855*10^-15+1.22909*10^-15 I	2.37185*10^-15
3.8	1.32874*10^-15+1.2158*10^-15 I	1.80104*10^-15
3.9	3.76478*10^-16+1.23905*10^-15 I	1.29499*10^-15
4.0	-1.60778*10^-15+5.64717*10^-17 I	1.60877*10^-15
4.1	-1.34203*10^-15-1.83588*10^-15 I	2.2741*10^-15
4.2	-6.37797*10^-16+2.3862*10^-15 I	2.46997*10^-15
4.3	-1.46162*10^-15+2.38067*10^-16 I	1.48088*10^-15
4.4	3.18899*10^-16+7.31363*10^-16 I	7.97865*10^-16
4.5	3.01182*10^-16+2.7959*10^-16 I	4.10952*10^-16
4.6	-1.06742*10^-15+1.25234*10^-15 I	1.64553*10^-15
4.7	1.00099*10^-15-1.02867*10^-15 I	1.43532*10^-15
4.8	-1.98869*10^-15+3.21058*10^-15 I	3.7766*10^-15
4.9	-6.06793*10^-16+1.46536*10^-15 I	1.58602*10^-15
5.0	-4.34057*10^-15+6.9376*10^-16 I	4.39566*10^-15
5.1	2.01969*10^-15-2.23077*10^-15 I	3.00923*10^-15
5.2	-2.43603*10^-16-5.53644*10^-17 I	2.49815*10^-16
5.3	-2.4316*10^-15+4.58417*10^-16 I	2.47444*10^-15
5.4	-6.2451*10^-16+6.26725*10^-16 I	8.84758*10^-16
5.5	-3.05611*10^-16+3.32574*10^-15 I	3.33975*10^-15
5.6	3.45474*10^-16+7.08664*10^-17 I	3.52667*10^-16
5.7	-1.52806*10^-15+1.93886*10^-15 I	2.46863*10^-15
5.8	1.05857*10^-15+2.86234*10^-15 I	3.05181*10^-15
5.9	1.02756*10^-15+2.14149*10^-15 I	2.37526*10^-15
6.0	-4.40257*10^-15+3.32408*10^-15 I	5.51653*10^-15
6.1	3.21113*10^-15+1.50591*10^-15 I	3.54671*10^-15
6.2	6.39569*10^-15-2.66856*10^-16 I	6.40126*10^-15

(continued)

Table 1.1 (continued)

n	A[n]	Abs [A[n]]
6.3	3.67619*10^-16+1.49152*10^-15 I	1.53615*10^-15
6.4	-5.5763*10^-15+5.20425*10^-17 I	5.57654*10^-15
6.5	-4.42915*10^-16+6.2606*10^-15 I	6.27625*10^-15
6.6	-2.44046*10^-15+1.8381*10^-16 I	2.44737*10^-15
6.7	3.67619*10^-15+2.83466*10^-15 I	4.64216*10^-15
6.8	2.46704*10^-15-1.34203*10^-15 I	2.80844*10^-15
6.9	8.28694*10^-15+1.19366*10^-15 I	8.37246*10^-15
7.0	-4.42915*10^-18-7.5827*10^-15 I	7.5827*10^-15
7.1	5.93063*10^-15-3.90429*10^-15 I	7.10042*10^-15
7.2	9.00889*10^-15+3.86*10^-15 I	9.80101*10^-15
7.3	1.34026*10^-14+5.50322*10^-15 I	1.44885*10^-14
7.4	7.44983*10^-15+7.00248*10^-15 I	1.02242*10^-14
7.5	6.46656*10^-16+1.18258*10^-14 I	1.18435*10^-14
7.6	1.06167*10^-14+5.97492*10^-15 I	1.21825*10^-14
7.7	-1.93554*10^-15-8.26036*10^-16 I	2.10443*10^-15
7.8	2.83864*10^-14-1.87132*10^-15 I	2.8448*10^-14
7.9	3.66999*10^-14+9.87479*10^-15 I	3.80052*10^-14
8.0	**1.99471+7.57429*10^-14 I**	**1.99471**
8.1	-1.7814*10^-14-8.28915*10^-15 I	1.96482*10^-14
8.2	-1.00276*10^-14+9.79285*10^-15 I	1.40162*10^-14
8.3	-1.93155*10^-14+7.03792*10^-15 I	2.05578*10^-14
8.4	-1.60335*10^-14-6.84968*10^-15 I	1.74354*10^-14
8.5	2.8258*10^-15-1.25046*10^-14 I	1.28199*10^-14
8.6	-8.69885*10^-15+3.58761*10^-15 I	9.40962*10^-15
8.7	-3.68505*10^-15-6.50421*10^-15 I	7.47558*10^-15
8.8	8.49511*10^-15-2.03298*10^-15 I	8.73498*10^-15
8.9	-1.35931*10^-14+3.87108*10^-15 I	1.41335*10^-14
9.0	8.60141*10^-15-2.83133*10^-15 I	9.05542*10^-15
9.1	-9.21263*10^-16+7.38671*10^-15 I	7.44394*10^-15

(continued)

Table 1.1   (continued)

n	A[n]	Abs [A[n]]
9.2	-5.41685*10^-15-4.8012*10^-15 I	7.23835*10^-15
9.3	5.31498*10^-16+1.00763*10^-16 I	5.40965*10^-16
9.4	-8.01676*10^-16+8.12195*10^-15 I	8.16142*10^-15
9.5	6.20081*10^-16+5.76011*10^-15 I	5.79339*10^-15
9.6	-8.26922*10^-15-3.87883*10^-15 I	9.13375*10^-15
9.7	2.88338*10^-15+4.29627*10^-16 I	2.91521*10^-15
9.8	-6.95376*10^-16+2.37402*10^-15 I	2.47377*10^-15
9.9	-4.9075*10^-15+9.50191*10^-16 I	4.99864*10^-15
10.0	-1.9754*10^-15+3.20943*10^-15 I	3.76864*10^-15

Key results are in bold face. $I = \sqrt{(-1)}$.

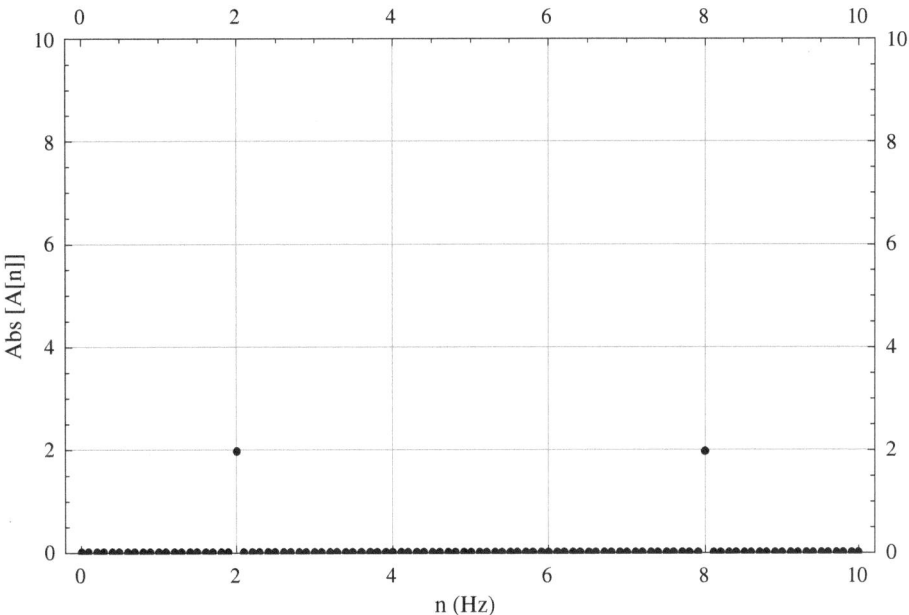

Fig. 1.1   Discrete Fourier transform for the numerical data for $y = \cos(2\pi\, 2\, t)$

Program number 1.1

```
h=N[10/100];
i=0;
Table[{i=i+1,t=t+h,y[i]=1*Cos[4*Pi*t]},{t,0,10-h,h}];
TableForm[%,TableSpacing->{3,3},
TableHeadings->{None,{"i","t","y[i]"}}]


i=0;
ListLinePlot[Table[{i=i+1;t=t+h,y[i]},{t,0,10-h,h}],
Frame->True,FrameLabel->{"t","y"}]


Z=N[Exp[-2*Pi*I/100]];


Table[{n1=n1+1;n=N[n1/10],A[n1]=h*(1/Sqrt[2*Pi])*(Sum[y[k]*(Z^(n1*k)),
{k,1,100,1}]),Ab[n1]=Abs[A[n1]]},{n1,0-1,100-1,1}];
TableForm[%,TableSpacing->{3,3},
TableHeadings->{None,{"n","A[n]","Abs [A[n]]"}}]


ListPlot[Table[{n1=n1+1;n=n1/10,Ab[n1]},{n1,0-1,100-1,1}],
Frame->True,FrameLabel->{"n (Hz)","Abs [A[n]]"},PlotStyle->{Black},
GridLines->Automatic,FrameTicks->All,PlotRange->{0,10}]
```

We find

(1) non-zero value of $A[n]$ for $n = \mathbf{2}$ demonstrating that indeed the frequency that is present is 2 Hz as in the function $y = \cos(2\pi\ \mathbf{2}\ t)$ taken.

(2) non-zero value of $A[n]$ for $n = 8$, which is unphysical but expected from Eq. (1.10) or (1.11): $A(n_{max} - n) = A^*(n)$ or $|A(n_{max} - n)|^2 = |A(n)|^2$

(3) $A[2] = 1.99471+1.42796*10^{\wedge}-14\ I$ is real and hence $A[10-2] = A[8] = 1.99471+7.57429*10^{\wedge}-14\ I$ is also real.

(4) that we need not use true period of the function $y = \cos(2\pi\ 2\ t)$ in calculating the frequency content of the data.

1.3   Discrete Fourier Transform: Example II

In this section, we take up the function $y = \sin(4\pi\ t) = \sin(2\pi\ 2\ t)$. Using Program number 1.2, we first turn the analytic function $y(t)$ into numerical data denoted by $y[i]$ for a discrete set of values of time in the interval 0–10 s. These discrete data are then fed to the rest of the program that calculates (the complex numbers) $A[n]$'s, and absolute values of them which constitute what is called power spectrum (Table 1.2 and Fig. 1.2).

Table 1.2 Discrete Fourier transform for the numerical data for $y = \sin(2\pi\ 2\ t)$

n	$A[n]$	$Abs\ [A[n]]$
0.0	-3.32729E-16	$3.32729*10^{-16}$
0.1	$-9.5281*10^{-16}-5.27346*10^{-16}$ I	$1.08901*10^{-15}$
0.2	$-1.56403*10^{-15}+7.54063*10^{-16}$ I	$1.73632*10^{-15}$
0.3	$1.19044*10^{-16}+9.85486*10^{-17}$ I	$1.54542*10^{-16}$
0.4	$-1.24513*10^{-15}-1.26895*10^{-15}$ I	$1.77781*10^{-15}$
0.5	$-1.25399*10^{-15}-1.84696*10^{-15}$ I	$2.23243*10^{-15}$
0.6	$7.34696*10^{-16}-6.57729*10^{-16}$ I	$9.86096*10^{-16}$
0.7	$-2.57433*10^{-16}+2.65749*10^{-17}$ I	$2.58801*10^{-16}$
0.8	$1.4119*10^{-16}-5.24854*10^{-16}$ I	$5.43513*10^{-16}$
0.9	$-1.44002*10^{-15}-1.15158*10^{-15}$ I	$1.84385*10^{-15}$
1.0	$8.3155*10^{-18}-1.72294*10^{-15}$ I	$1.72296*10^{-15}$
1.1	$1.74011*10^{-15}+3.23328*10^{-16}$ I	$1.7699*10^{-15}$
1.2	$2.08116*10^{-15}+1.59449*10^{-16}$ I	$2.08726*10^{-15}$
1.3	$1.76669*10^{-15}-2.17028*10^{-16}$ I	$1.77997*10^{-15}$
1.4	$1.74454*10^{-15}-1.02313*10^{-15}$ I	$2.02243*10^{-15}$
1.5	$2.08559*10^{-15}+1.46162*10^{-16}$ I	$2.0907*10^{-15}$
1.6	$7.657*10^{-16}-2.70621*10^{-15}$ I	$2.81245*10^{-15}$
1.7	$-2.64253*10^{-15}-3.06497*10^{-15}$ I	$4.04685*10^{-15}$
1.8	$-2.02023*10^{-15}-5.15996*10^{-15}$ I	$5.54135*10^{-15}$
1.9	$4.12521*10^{-15}-5.97492*10^{-15}$ I	$7.26065*10^{-15}$
2.0	$1.44119*10^{-14}$-**1.99471 I**	**1.99471**
2.1	$-7.66786*10^{-16}-1.48376*10^{-15}$ I	$1.67019*10^{-15}$
2.2	$1.56184*10^{-15}+2.64863*10^{-15}$ I	$3.07483*10^{-15}$
2.3	$-3.35673*10^{-15}+4.11911*10^{-15}$ I	$5.31363*10^{-15}$
2.4	$3.65139*10^{-16}+4.42915*10^{-18}$ I	$3.65166*10^{-16}$
2.5	$6.55256*10^{-16}+3.90208*10^{-15}$ I	$3.95671*10^{-15}$

(continued)

Table 1.2 (continued)

n	A[n]	Abs [A[n]]
2.6	-1.71878*10^-15-2.70621*10^-15 I	3.2059*10^-15
2.7	8.0667*10^-16+1.36861*10^-15 I	1.58865*10^-15
2.8	1.18868*10^-15+2.0507*10^-15 I	2.3703*10^-15
2.9	1.161*10^-15+1.16044*10^-15 I	1.64151*10^-15
3.0	-2.66911*10^-15+5.19539*10^-15 I	5.84091*10^-15
3.1	-1.44223*10^-15-7.61814*10^-16 I	1.63107*10^-15
3.2	5.02166*10^-16-5.80219*10^-16 I	7.67349*10^-16
3.3	-5.51972*10^-16+3.24214*10^-15 I	3.28879*10^-15
3.4	2.94927*10^-15+1.81595*10^-16 I	2.95486*10^-15
3.5	3.84793*10^-16+1.72737*10^-15 I	1.76971*10^-15
3.6	-1.16098*10^-15-2.28544*10^-15 I	2.56342*10^-15
3.7	1.36364*10^-15+2.97639*10^-15 I	3.27389*10^-15
3.8	2.21846*10^-15+1.34203*10^-15 I	2.5928*10^-15
3.9	-2.24612*10^-15+5.54087*10^-15 I	5.97882*10^-15
4.0	9.47295*10^-16+4.1634*10^-16 I	1.03475*10^-15
4.1	-4.8416*10^-15+2.25887*10^-15 I	5.34262*10^-15
4.2	-2.54287*10^-15+2.88781*10^-15 I	3.84781*10^-15
4.3	-2.01581*10^-15-2.13706*10^-15 I	2.93777*10^-15
4.4	-1.93608*10^-15-6.09008*10^-16 I	2.02961*10^-15
4.5	-6.51628*10^-16-4.42915*10^-18 I	6.51643*10^-16
4.6	1.85527*10^-15+4.42915*10^-18 I	1.85528*10^-15
4.7	3.13927*10^-16+7.3081*10^-17 I	3.22321*10^-16
4.8	-3.51286*10^-15+8.30465*10^-18 I	3.51287*10^-15
4.9	3.93195*10^-17+1.17345*10^-15 I	1.17411*10^-15
5.0	1.37692*10^-15-6.99818*10^-17 I	1.3787*10^-15
5.1	1.71354*10^-15+1.12666*10^-16 I	1.71724*10^-15
5.2	-2.01138*10^-15-8.68667*10^-16 I	2.19094*10^-15
5.3	-1.56403*10^-15+1.53691*10^-15 I	2.19278*10^-15
5.4	9.16291*10^-16+3.21113*10^-17 I	9.16854*10^-16
5.5	3.04612*10^-17-1.25566*10^-15 I	1.25603*10^-15
5.6	-1.99809*10^-15-1.2313*10^-15 I	2.34701*10^-15
5.7	-4.97891*10^-15+2.22343*10^-15 I	5.45281*10^-15

(continued)

Table 1.2 (continued)

n	A[n]	Abs [A[n]]
5.8	-4.30568*10^-15+1.07628*10^-15 I	4.43816*10^-15
5.9	-1.58175*10^-15+8.63684*10^-16 I	1.80219*10^-15
6.0	-3.3977*10^-15-1.48819*10^-15 I	3.70932*10^-15
6.1	-1.15655*10^-15+1.2003*10^-15 I	1.66683*10^-15
6.2	1.77998*10^-15+4.75691*10^-15 I	5.07902*10^-15
6.3	-2.09996*10^-15+3.56104*10^-15 I	4.1341*10^-15
6.4	-2.57831*10^-15+1.36418*10^-15 I	2.91696*10^-15
6.5	-2.44986*10^-15-4.38486*10^-15 I	5.02283*10^-15
6.6	-9.43952*10^-16-2.57776*10^-15 I	2.74516*10^-15
6.7	4.2687*10^-16+3.95523*10^-15 I	3.9782*10^-15
6.8	5.21921*10^-15+2.21457*10^-15 I	5.66961*10^-15
6.9	-2.08889*10^-15+7.43211*10^-15 I	7.72009*10^-15
7.0	4.14736*10^-15-6.32925*10^-15 I	7.56703*10^-15
7.1	4.83166*10^-15+1.17195*10^-14 I	1.26764*10^-14
7.2	2.83522*10^-15+8.21607*10^-15 I	8.69151*10^-15
7.3	-2.84572*10^-15+1.19941*10^-14 I	1.23271*10^-14
7.4	-6.93908*10^-15+9.75299*10^-15 I	1.19696*10^-14
7.5	-1.11766*10^-14+5.09795*10^-15 I	1.22844*10^-14
7.6	-5.4149*10^-15+6.89176*10^-15 I	8.76456*10^-15
7.7	6.52747*10^-15+1.70522*10^-15 I	6.74653*10^-15
7.8	4.93131*10^-15+2.67875*10^-14 I	2.72376*10^-14
7.9	-1.47197*10^-14+3.63412*10^-14 I	3.92091*10^-14
8.0	-7.55862*10^-14+**1.99471 I**	**1.99471**
8.1	3.73102*10^-15-1.42176*10^-14 I	1.4699*10^-14
8.2	-1.38948*10^-14-3.06054*10^-15 I	1.42279*10^-14
8.3	-5.41075*10^-15-2.18003*10^-14 I	2.24617*10^-14
8.4	5.44288*10^-15-1.56393*10^-14 I	1.65594*10^-14
8.5	1.2627*10^-14+2.88781*10^-15 I	1.2953*10^-14
8.6	-1.6885*10^-16-6.28939*10^-15 I	6.29166*10^-15
8.7	5.7042*10^-15-5.13781*10^-15 I	7.67692*10^-15
8.8	1.7534*10^-15+8.25593*10^-15 I	8.44007*10^-15
8.9	-5.94003*10^-15-7.76873*10^-15 I	9.77942*10^-15
9.0	7.81248*10^-15+1.11039*10^-14 I	1.35769*10^-14
9.1	-6.21021*10^-15-7.97247*10^-16 I	6.26117*10^-15

(continued)

Table 1.2 (continued)

n	$A[n]$	$Abs\,[A[n]]$
9.2	-1.59992*10^-16-8.48625*10^-15 I	8.48776*10^-15
9.3	2.8164*10^-15-5.439*10^-15 I	6.12493*10^-15
9.4	3.49848*10^-15+2.11935*10^-15 I	4.09036*10^-15
9.5	-5.45328*10^-16-4.05932*10^-15 I	4.09578*10^-15
9.6	5.69534*10^-15-5.27733*10^-15 I	7.76448*10^-15
9.7	2.36019*10^-15+3.01957*10^-15 I	3.83254*10^-15
9.8	3.9104*10^-15-4.32672*10^-15 I	5.83196*10^-15
9.9	4.37989*10^-15+3.58761*10^-16 I	4.39455*10^-15
10.0	7.52413*10^-16-6.61896*10^-15 I	6.66159*10^-15

Key results are in bold face. $I = \sqrt{(-1)}$.

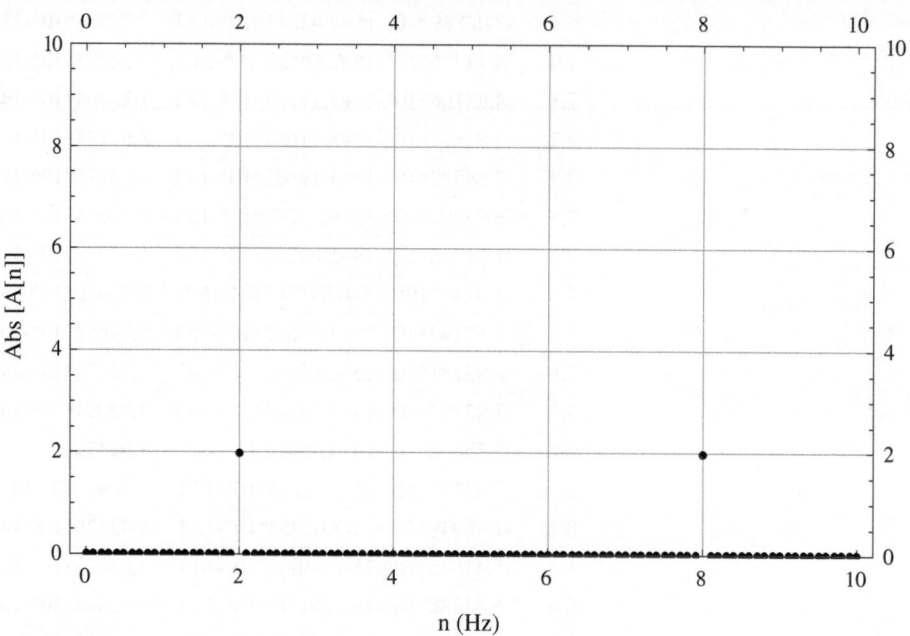

Fig. 1.2 Discrete Fourier transform for the numerical data for $y = \sin(2\pi\,2\,t)$

Program number 1.2

```
h=N[10/100];
i=0;
Table[{i=i+1,t=t+h,y[i]=1*Sin[4*Pi*t]},{t,0,10-h,h}];
TableForm[%,TableSpacing->{3,3},
TableHeadings->{None,{"i","t","y[i]"}}]


i=0;
ListLinePlot[Table[{i=i+1;t=t+h,y[i]},{t,0,10-h,h}],
Frame->True,FrameLabel->{"t","y"}]


Z=N[Exp[-2*Pi*I/100]];


Table[{n1=n1+1;n=N[n1/10],A[n1]=h*(1/Sqrt[2*Pi])*(Sum[y[k]*(Z^(n1*k)),
{k,1,100,1}]),Ab[n1]=Abs[A[n1]]},{n1,0-1,100-1,1}];
TableForm[%,TableSpacing->{3,3},
TableHeadings->{None,{"n","A[n]","Abs [A[n]]"}}]


ListPlot[Table[{n1=n1+1;n=n1/10,Ab[n1]},{n1,0-1,100-1,1}],
Frame->True,FrameLabel->{"n (Hz)","Abs [A[n]]"},PlotStyle->{Black},
GridLines->Automatic,FrameTicks->All,PlotRange->{0,10}]
```

We find

(1) non-zero value of $A[n]$ for $n = 2$ demonstrating that indeed the frequency that is present is 2 Hz as in the function $y = \sin(2\pi \ 2 \ t)$ taken.

(2) non-zero value of $A[n]$ for $n = 8$, which is unphysical but expected from Eq. (1.10) or (1.11): $A(n_{max} - n) = A^*(n)$ or $|A(n_{max} - n)|^2 = |A(n)|^2$

(3) that $A[2] = 1.44119*10\hat{\ }-14-1.99471 \ I$ and $A[10-2] = A[8] = -7.55862*10\hat{\ }-14+1.99471 \ I$ are complex conjugate of each other, as expected from Eq. (1.10): $A(n_{max} - n) = A^*(n)$.

(4) that we need not use true period of the function $y = \sin(2\pi \ 2 \ t)$ in calculating the frequency content of the data.

(5) that $A[2]$ is real number for cosine function (see Table 1.1) while $A[2]$ is purely imaginary number having no real part for sine function (see Table 1.2).

1.4   Discrete Fourier Transform: Example III

In this section, we take up the function $y = \sin(4\pi\, t + \pi/6) = \sin(2\pi\, 2\, t + \pi/6)$. Using Program number 1.3, we first turn the analytic function $y(t)$ into numerical data denoted by $y[i]$ for a discrete set of values of time t in the interval 0–10 s. These discrete data are then fed to the rest of the program that calculates (the complex numbers) $A[n]$'s, and absolute values of them which constitute what is called power spectrum (Table 1.3 and Fig. 1.3).

Table 1.3 Discrete Fourier transform for the numerical data for $y = \sin(2\pi\, 2\, t + \pi/6)$

n	A[n]	Abs [A[n]]
0.0	2.07063*10^-15	2.07063*10^-15
0.1	-3.65405*10^-16+2.23395*10^-16 I	4.28283*10^-16
0.2	-1.15822*10^-15+2.0208*10^-15 I	2.32919*10^-15
0.3	-1.77166*10^-17+4.5731*10^-16 I	4.57653*10^-16
0.4	-1.51477*10^-15-1.07518*10^-15 I	1.85756*10^-15
0.5	-1.27338*10^-15-1.81374*10^-15 I	2.21611*10^-15
0.6	1.16044*10^-15-7.08664*10^-16 I	1.35971*10^-15
0.7	-4.11911*10^-16-1.10729*10^-17 I	4.1206*10^-16
0.8	1.72737*10^-16-4.45129*10^-16 I	4.77471*10^-16
0.9	-1.63214*10^-15-1.8082*10^-15 I	2.43587*10^-15
1.0	1.67643*10^-15-1.76945*10^-15 I	2.43749*10^-15
1.1	1.22245*10^-15+1.08625*10^-15 I	1.63533*10^-15
1.2	1.45276*10^-15+1.37304*10^-15 I	1.99894*10^-15
1.3	1.90896*10^-15+1.28335*10^-15 I	2.30024*10^-15
1.4	1.83145*10^-15-1.24016*10^-16 I	1.83565*10^-15
1.5	1.99533*10^-15+8.60362*10^-16 I	2.17292*10^-15
1.6	1.74066*10^-15-1.00099*10^-15 I	2.00795*10^-15
1.7	6.2451*10^-16-3.01293*10^-15 I	3.07697*10^-15
1.8	1.49705*10^-15-4.40479*10^-15 I	4.65224*10^-15
1.9	4.8477*10^-15-4.6196*10^-15 I	6.69634*10^-15
2.0	**0.997356-1.72747 I**	**1.99471**
2.1	-3.82678*10^-15-1.01649*10^-15 I	3.95949*10^-15
2.2	-8.23822*10^-16+3.66955*10^-15 I	3.76089*10^-15
2.3	-4.15233*10^-15+2.3098*10^-15 I	4.75153*10^-15

(continued)

Table 1.3 (continued)

n	A[n]	Abs [A[n]]
2.4	1.26895*10^-15+2.56891*10^-16 I	1.29469*10^-15
2.5	-2.24336*10^-15+4.08921*10^-15 I	4.66415*10^-15
2.6	-4.71704*10^-16-3.31522*10^-15 I	3.34861*10^-15
2.7	-7.75101*10^-17+1.73069*10^-15 I	1.73242*10^-15
2.8	-1.79381*10^-16+2.35188*10^-15 I	2.35871*10^-15
2.9	4.25198*10^-16+1.51366*10^-15 I	1.57225*10^-15
3.0	-5.35927*10^-15+4.77241*10^-15 I	7.17619*10^-15
3.1	-2.50247*10^-16-1.43726*10^-15 I	1.45888*10^-15
3.2	1.37304*10^-16-6.20081*10^-16 I	6.351*10^-16
3.3	-1.11172*10^-15+2.48918*10^-15 I	2.72616*10^-15
3.4	1.99976*10^-15+7.64028*10^-16 I	2.14074*10^-15
3.5	-2.3253*10^-16+2.25887*10^-15 I	2.2708*10^-15
3.6	-2.65749*10^-17-2.55562*10^-15 I	2.55576*10^-15
3.7	1.41733*10^-16+3.30415*10^-15 I	3.30718*10^-15
3.8	2.60434*10^-15+1.85581*10^-15 I	3.19791*10^-15
3.9	-1.67865*10^-15+5.44564*10^-15 I	5.69849*10^-15
4.0	4.42915*10^-18+4.27413*10^-16 I	4.27436*10^-16
4.1	-4.84106*10^-15+1.11836*10^-15 I	4.96856*10^-15
4.2	-2.41167*10^-15+3.74927*10^-15 I	4.45794*10^-15
4.3	-2.38953*10^-15-1.76834*10^-15 I	2.97268*10^-15
4.4	-1.5037*10^-15-1.91561*10^-16 I	1.51585*10^-15
4.5	-4.05267*10^-16+1.77166*10^-16 I	4.423*10^-16
4.6	1.16487*10^-15+6.64372*10^-16 I	1.34101*10^-15
4.7	7.66243*10^-16-5.08245*10^-16 I	9.19479*10^-16
4.8	-4.00395*10^-15+1.60114*10^-15 I	4.31222*10^-15
4.9	-2.34745*10^-16+1.78329*10^-15 I	1.79867*10^-15
5.0	-9.1019*10^-16+2.86274*10^-16 I	9.54148*10^-16
5.1	2.50468*10^-15-1.09704*10^-15 I	2.7344*10^-15
5.2	-1.85138*10^-15-7.69011*10^-16 I	2.00474*10^-15
5.3	-2.57334*10^-15+1.6266*10^-15 I	3.04432*10^-15
5.4	5.7136*10^-16+3.47688*10^-16 I	6.68835*10^-16
5.5	-1.10729*10^-16+5.2264*10^-16 I	5.34241*10^-16

(continued)

Table 1.3 (continued)

n	A[n]	Abs [A[n]]
5.6	-1.5347*10^-15-9.98773*10^-16 I	1.83108*10^-15
5.7	-5.00051*10^-15+2.92102*10^-15 I	5.79115*10^-15
5.8	-3.12255*10^-15+2.28323*10^-15 I	3.86826*10^-15
5.9	-8.30465*10^-16+1.77498*10^-15 I	1.95965*10^-15
6.0	-5.11788*10^-15+3.32186*10^-16 I	5.12865*10^-15
6.1	6.33368*10^-16+1.77277*10^-15 I	1.88251*10^-15
6.2	4.79234*10^-15+3.88436*10^-15 I	6.16886*10^-15
6.3	-1.65207*10^-15+3.73599*10^-15 I	4.08497*10^-15
6.4	-5.06252*10^-15+1.2313*10^-15 I	5.2101*10^-15
6.5	-2.29208*10^-15-6.80982*10^-16 I	2.39111*10^-15
6.6	-2.04184*10^-15-2.2633*10^-15 I	3.04821*10^-15
6.7	2.12378*10^-15+4.76687*10^-15 I	5.21857*10^-15
6.8	5.7291*10^-15+1.29331*10^-15 I	5.87327*10^-15
6.9	2.4316*10^-15+6.99141*10^-15 I	7.4022*10^-15
7.0	3.56104*10^-15-9.43409*10^-15 I	1.00838*10^-14
7.1	6.98034*10^-15+8.14078*10^-15 I	1.07237*10^-14
7.2	6.95155*10^-15+9.10633*10^-15 I	1.14564*10^-14
7.3	4.30513*10^-15+1.29996*10^-14 I	1.36939*10^-14
7.4	-2.2943*10^-15+1.16664*10^-14 I	1.18898*10^-14
7.5	-9.63783*10^-15+1.02147*10^-14 I	1.40438*10^-14
7.6	6.88733*10^-16+8.93359*10^-15 I	8.9601*10^-15
7.7	4.97393*10^-15+6.3669*10^-16 I	5.01452*10^-15
7.8	1.80687*10^-14+2.13441*10^-14 I	2.79651*10^-14
7.9	4.0571*10^-15+3.56901*10^-14 I	3.59199*10^-14
8.0	**0.997356+1.72747 I**	**1.99471**
8.1	-5.7291*10^-15-1.46561*10^-14 I	1.5736*10^-14
8.2	-1.64211*10^-14+3.11812*10^-15 I	1.67145*10^-14
8.3	-1.36506*10^-14-1.544*10^-14 I	2.06091*10^-14
8.4	-3.35508*10^-15-1.70079*10^-14 I	1.73357*10^-14
8.5	1.23706*10^-14-3.47688*10^-15 I	1.28499*10^-14
8.6	-4.22319*10^-15-3.43923*10^-15 I	5.44644*10^-15
8.7	3.28643*10^-15-7.84734*10^-15 I	8.50773*10^-15
8.8	5.77782*10^-15+6.07458*10^-15 I	8.38354*10^-15
8.9	-1.1941*10^-14-4.60853*10^-15 I	1.27994*10^-14

(continued)

Table 1.3 (continued)

n	A[n]	Abs [A[n]]
9.0	1.1239*10^-14+8.34009*10^-15 I	1.39954*10^-14
9.1	-5.72468*10^-15+2.9332*10^-15 I	6.43239*10^-15
9.2	-2.88338*10^-15-9.74413*10^-15 I	1.01618*10^-14
9.3	2.69957*10^-15-4.48451*10^-15 I	5.23436*10^-15
9.4	2.8258*10^-15+6.01811*10^-15 I	6.64851*10^-15
9.5	-5.53644*10^-17-6.93162*10^-16 I	6.95369*10^-16
9.6	8.19393*10^-16-6.48317*10^-15 I	6.53474*10^-15
9.7	3.62526*10^-15+2.99743*10^-15 I	4.70394*10^-15
9.8	3.26871*10^-15-2.50911*10^-15 I	4.12069*10^-15
9.9	1.42397*10^-15+6.94823*10^-16 I	1.58445*10^-15
10.0	-3.47688*10^-16-4.12748*10^-15 I	4.14209*10^-15

Key results are in bold face. $I = \sqrt{(-1)}$.

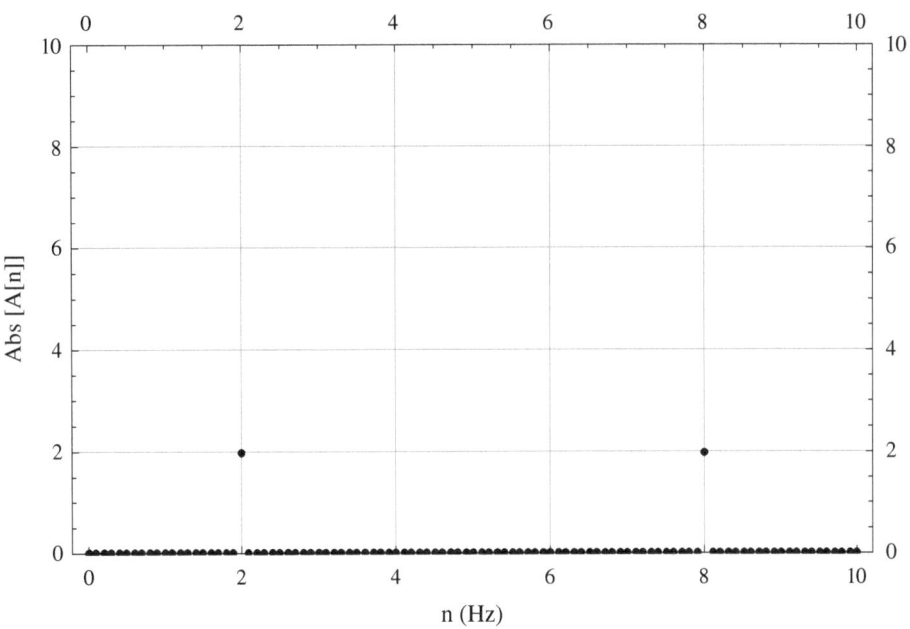

Fig. 1.3 Discrete Fourier transform for the numerical data for $y = \sin(2\pi\, 2\, t + \pi/6)$

Program number 1.3

```
h=N[10/100];
i=0;
Table[{i=i+1,t=t+h,y[i]=1*Sin[4*Pi*t+Pi/6]},{t,0,10-h,h}];
TableForm[%,TableSpacing->{3,3},
TableHeadings->{None,{"i","t","y[i]"}}]


i=0;
ListLinePlot[Table[{i=i+1;t=t+h,y[i]},{t,0,10-h,h}],
Frame->True,FrameLabel->{"t","y"}]


Z=N[Exp[-2*Pi*I/100]];


Table[{n1=n1+1;n=N[n1/10],A[n1]=h*(1/Sqrt[2*Pi])*(Sum[y[k]*(Z^(n1*k)),
{k,1,100,1}]),Ab[n1]=Abs[A[n1]]},{n1,0-1,100-1,1}];
TableForm[%,TableSpacing->{3,3},
TableHeadings->{None,{"n","A[n]","Abs [A[n]]"}}]


ListPlot[Table[{n1=n1+1;n=n1/10,Ab[n1]},{n1,0-1,100-1,1}],
Frame->True,FrameLabel->{"n (Hz)","Abs [A[n]]"},PlotStyle->{Black},
GridLines->Automatic,FrameTicks->All,PlotRange->{0,10}]
```

We find

(1) non-zero value of $A[n]$ for $n = 2$ demonstrating that indeed the frequency that is present is 2 Hz as in the function $y = \sin(2\pi\ 2\ t + \pi/6)$ taken.
(2) non-zero value of $A[n]$ for $n = 8$, which is unphysical but expected from Eq. (1.10) or (1.11): $A(n_{max} - n) = A^*(n)$ or $|A(n_{max} - n)|^2 = |A(n)|^2$
(3) that $A[2] = 0.997356-1.72747\ I$ and $A[10-2] = A[8] = 0.997356+1.72747\ I$ are complex conjugate of each other, as expected from Eq. (1.10): $A(n_{max} - n) = A^*(n)$.
(4) that we need not use true period of the function $y = \sin(2\pi\ 2\ t + \pi/6)$ in calculating the frequency content of the data.
(5) that $A[2]$ has both real and imaginary parts as $y = \sin(2\pi\ 2\ t + \pi/6)$ has both sine and cosine terms in it.

1.5   Discrete Fourier Transform: Example IV

In this section, we take up the function $y = 3/2 + 3 \cos(2\pi\ t) + 4 \sin(6\pi\ t) = 3/2 + 3 \cos(2\pi\ 1\ t) + 4 \sin(2\pi\ 3\ t)$. Using Program number 1.4, we first turn the analytic function $y(t)$ into numerical data denoted by $y[i]$ for a discrete set of values of time t in the interval 0–10 s. These discrete data are then fed to the rest of the program that calculates (the complex numbers) $A[n]$'s, and absolute value of them which constitute what is called power spectrum (Table 1.4 and Fig. 1.4).

Table 1.4 Discrete Fourier transform for the numerical data for $y = 3/2 + 3 \cos(2\pi\ 1\ t) + 4 \sin(2\pi\ 3\ t)$

n	$A[n]$	$Abs\ [A[n]]$
0.0	**5.98413**	**5.98413**
0.1	-1.20473*10^-15+8.28247*10^-15 I	8.36963*10^-15
0.2	-8.04333*10^-15+2.49015*10^-15 I	8.41998*10^-15
0.3	-5.70474*10^-15+6.49645*10^-15 I	8.64569*10^-15
0.4	-1.27559*10^-15-1.39837*10^-15 I	1.89277*10^-15
0.5	-8.18507*10^-15-6.56068*10^-17 I	8.18533*10^-15
0.6	-1.14804*10^-14+7.44263*10^-15 I	1.36818*10^-14
0.7	-7.83074*10^-15+3.59149*10^-15 I	8.61506*10^-15
0.8	-1.84253*10^-14+1.32985*10^-15 I	1.84732*10^-14
0.9	-7.7953*10^-15-1.17616*10^-14 I	1.41104*10^-14
1.0	**5.98413+2.06847*10^-14 I**	**5.98413**
1.1	-7.22837*10^-15-1.99063*10^-14 I	2.1178*10^-14
1.2	-1.2189*10^-14-5.95776*10^-15 I	1.35671*10^-14
1.3	8.04333*10^-15-1.16232*10^-14 I	1.41348*10^-14
1.4	-9.6024*10^-15-1.51145*10^-15 I	9.72062*10^-15
1.5	-9.2835*10^-15-1.76502*10^-15 I	9.44979*10^-15
1.6	-3.33072*10^-15-6.47929*10^-15 I	7.28525*10^-15
1.7	-4.53545*10^-15+8.39988*10^-15 I	9.54611*10^-15
1.8	1.13386*10^-14-1.53193*10^-15 I	1.14416*10^-14
1.9	-1.56969*10^-14-1.03365*10^-15 I	1.57309*10^-14
2.0	-1.38898*10^-14-2.30814*10^-15 I	1.40803*10^-14
2.1	-1.24725*10^-14-2.93099*10^-15 I	1.28122*10^-14
2.2	-5.84648*10^-15-5.44509*10^-15 I	7.98938*10^-15
2.3	-8.2205*10^-15-1.97258*10^-14 I	2.13701*10^-14

(continued)

Table 1.4 (continued)

n	A[n]	Abs [A[n]]
2.4	4.96065*10^-16-8.87989*10^-15 I	8.89374*10^-15
2.5	2.87009*10^-15-1.9563*10^-14 I	1.97724*10^-14
2.6	2.05513*10^-15+3.66955*10^-15 I	4.20585*10^-15
2.7	2.43072*10^-14-2.06426*10^-14 I	3.18897*10^-14
2.8	2.2004*10^-14+1.00475*10^-14 I	2.41894*10^-14
2.9	-5.70474*10^-15+2.01476*10^-14 I	2.09397*10^-14
3.0	7.54018*10^-14-**7.97885 I**	**7.97885**
3.1	-9.92129*10^-16+4.31377*10^-14 I	4.31491*10^-14
3.2	-4.02167*10^-14+2.46593*10^-14 I	4.71748*10^-14
3.3	6.73231*10^-16+1.15844*10^-14 I	1.1604*10^-14
3.4	-8.752*10^-15-3.18157*10^-14 I	3.29975*10^-14
3.5	2.97993*10^-14+1.1121*10^-14 I	3.18069*10^-14
3.6	1.33583*10^-14+2.51703*10^-14 I	2.84954*10^-14
3.7	-1.35*10^-14+1.96305*10^-14 I	2.38246*10^-14
3.8	4.05356*10^-14+7.4908*10^-15 I	4.12219*10^-14
3.9	-9.14176*10^-15+9.70427*10^-15 I	1.33321*10^-14
4.0	-8.36223*10^-15+1.03753*10^-15 I	8.42635*10^-15
4.1	8.89373*10^-15+2.05075*10^-14 I	2.2353*10^-14
4.2	-7.29924*10^-15+4.36548*10^-16 I	7.31228*10^-15
4.3	-3.18899*10^-16+7.71779*10^-15 I	7.72438*10^-15
4.4	4.81891*10^-15+2.86704*10^-15 I	5.60731*10^-15
4.5	-4.53545*10^-15+1.38148*10^-14 I	1.45402*10^-14
4.6	1.50237*10^-14+7.11224*10^-15 I	1.66221*10^-14
4.7	-1.47048*10^-14+2.35992*10^-14 I	2.78056*10^-14
4.8	-2.84529*10^-14+2.08592*10^-15 I	2.85292*10^-14
4.9	-1.13386*10^-14-8.39165*10^-15 I	1.41062*10^-14
5.0	-7.1575*10^-15+1.531*10^-14 I	1.69005*10^-14
5.1	5.95278*10^-15+5.46187*10^-15 I	8.07883*10^-15
5.2	-1.55906*10^-14+4.54092*10^-15 I	1.62384*10^-14
5.3	-3.08269*10^-15+1.12859*10^-14 I	1.16993*10^-14
5.4	-1.33583*10^-14+1.63103*10^-15 I	1.34575*10^-14
5.5	-3.27403*10^-14-5.75845*10^-15 I	3.32428*10^-14
5.6	-9.00003*10^-15+6.97591*10^-15 I	1.1387*10^-14
5.7	-1.02402*10^-14+2.52434*10^-15 I	1.05467*10^-14

(continued)

Table 1.4 (continued)

n	A[n]	Abs [A[n]]
5.8	-3.90474*10^-14+1.4*10^-14 I	4.14813*10^-14
5.9	-3.82678*10^-15+3.99465*10^-14 I	4.01294*10^-14
6.0	-2.0445*10^-14+1.10552*10^-14 I	2.32425*10^-14
6.1	-9.92129*10^-16+3.32806*10^-14 I	3.32954*10^-14
6.2	6.41341*10^-15+5.29383*10^-14 I	5.33254*10^-14
6.3	1.56615*10^-14+4.09846*10^-14 I	4.3875*10^-14
6.4	4.73033*10^-14+4.92228*10^-14 I	6.82678*10^-14
6.5	2.05867*10^-14+3.60267*10^-14 I	4.14938*10^-14
6.6	-8.82286*10^-15+6.6992*10^-14 I	6.75705*10^-14
6.7	3.57521*10^-14+6.73939*10^-14 I	7.62899*10^-14
6.8	-2.50158*10^-14+1.15871*10^-13 I	1.18541*10^-13
6.9	-3.43348*10^-14+9.06652*10^-14 I	9.69488*10^-14
7.0	-2.69895*10^-13+**7.97885 I**	**7.97885**
7.1	4.75159*10^-14-9.62604*10^-14 I	1.07349*10^-13
7.2	4.20238*10^-14+1.17937*10^-14 I	4.36473*10^-14
7.3	5.06695*10^-15+9.60184*10^-15 I	1.08568*10^-14
7.4	-3.21379*10^-14-1.76164*10^-14 I	3.66494*10^-14
7.5	9.3898*10^-15-4.27585*10^-14 I	4.37773*10^-14
7.6	1.75749*10^-14-1.66924*10^-15 I	1.7654*10^-14
7.7	-1.6193*10^-14+5.21848*10^-14 I	5.46394*10^-14
7.8	-6.19018*10^-14-5.04901*10^-14 I	7.98817*10^-14
7.9	1.21536*10^-14+1.88554*10^-14 I	2.2433*10^-14
8.0	7.61814*10^-15+6.20524*10^-15 I	9.82553*10^-15
8.1	1.25079*10^-14+6.82587*10^-15 I	1.42492*10^-14
8.2	2.6504*10^-14+5.16217*10^-15 I	2.70021*10^-14
8.3	-7.29924*10^-15+2.01781*10^-14 I	2.14577*10^-14
8.4	-2.87009*10^-14-4.73792*10^-14 I	5.53943*10^-14
8.5	1.69725*10^-14+3.16263*10^-14 I	3.58928*10^-14
8.6	8.0079*10^-15+2.80182*10^-14 I	2.91402*10^-14
8.7	8.57483*10^-15+3.63666*10^-14 I	3.73639*10^-14
8.8	4.92521*10^-15-3.46027*10^-16 I	4.93735*10^-15
8.9	1.08461*10^-13+3.57853*10^-14 I	1.14212*10^-13
9.0	**5.98413**+2.69275*10^-13 I	**5.98413**
9.1	-3.87285*10^-14+2.15766*10^-14 I	4.43333*10^-14

(continued)

Table 1.4 (continued)

n	A[n]	Abs [A[n]]
9.2	-7.31341*10^-14+2.56531*10^-14 I	7.75028*10^-14
9.3	-4.23072*10^-14-8.45104*10^-14 I	9.45088*10^-14
9.4	7.3205*10^-14-9.42523*10^-15 I	7.38092*10^-14
9.5	1.58386*10^-14+5.53469*10^-14 I	5.75686*10^-14
9.6	-9.46066*10^-15-3.29857*10^-14 I	3.43156*10^-14
9.7	1.71497*10^-14+7.93355*10^-14 I	8.11679*10^-14
9.8	-3.51852*10^-14-2.26623*10^-14 I	4.18518*10^-14
9.9	1.1863*10^-13-1.80499*10^-14 I	1.19996*10^-13
10.0	**5.98413+2.78802*10^-13 I**	**5.98413**

Key results are in bold face. $I = \sqrt{(-1)}$.

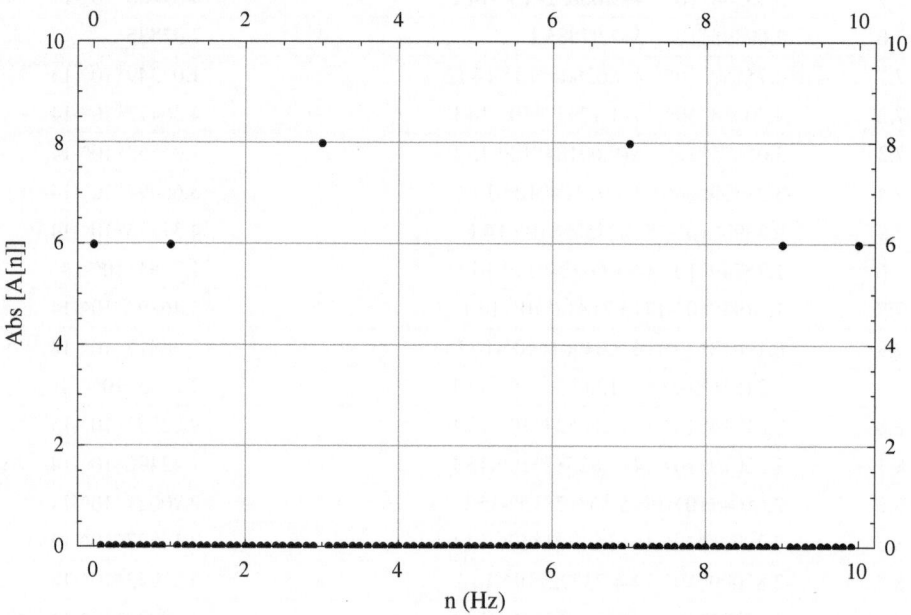

Fig. 1.4 Discrete Fourier transform for the numerical data for $y = 3/2 + 3 \cos(2\pi \, 1 \, t) + 4 \sin(2\pi \, 3 \, t)$

Program number 1.4

```
h=N[10/100];
i=0;
Table[{i=i+1,t=t+h, y[i]=3/2+3*Cos[2*Pi*1*t]+4*Sin[2*Pi*3*t]},{t,0,10-h,h}];
TableForm[%,TableSpacing->{3,3},
TableHeadings->{None,{"i","t","y[i]"}}]
```

```
i=0;
ListLinePlot[Table[{i=i+1;t=t+h,y[i]},{t,0,10-h,h}],
Frame->True,FrameLabel->{"t","y"}]
```

```
Z=N[Exp[-2*Pi*I/100]];
```

```
Table[{n1=n1+1;n=N[n1/10],A[n1]=h*(1/Sqrt[2*Pi])*(Sum[y[k]*(Z^(n1*k)),
{k,1,100,1}]),Ab[n1]=Abs[A[n1]]},{n1,0-1,100-1,1}];
TableForm[%,TableSpacing->{3,3},
TableHeadings->{None,{"n","A[n]","Abs [A[n]]"}}]
```

```
ListPlot[Table[{n1=n1+1;n=n1/10,Ab[n1]},{n1,0-1,100-1,1}],
Frame->True,FrameLabel->{"n (Hz)","Abs [A[n]]"},PlotStyle->{Black},
GridLines->Automatic,FrameTicks->All,PlotRange->{0,10}]
```

We find

(1) non-zero value of $A[n]$ for $n = \mathbf{0}$ demonstrating that indeed we have a "dc component" present in the function $y = \mathbf{3/2} + 3\cos(2\pi\,1\,t) + 4\sin(2\pi\,3\,t)$ taken.

(2) non-zero value of $A[n]$ for $n = \mathbf{1\ and\ 3}$ demonstrating that indeed the frequencies that are present are 1 and 3 Hz in the function $y = 3/2 + 3\cos(2\pi\,\mathbf{1}\,t) + 4\sin(2\pi\,\mathbf{3}\,t)$ taken.

(3) non-zero value of $A[n]$ for $n = 7, 9$ and 10, which are unphysical but expected from Eq. (1.10) or (1.11): $A(n_{max} - n) = A^*(n)$ or $|A(n_{max} - n)|^2 = |A(n)|^2$.

(4) $A[0] = 5.98413$ and hence $A[0]$ and $A[10]$ are real.

(5) that $A[1] = 5.98413+2.06847*10^{\wedge}-14\ I$ is real as it corresponds to the cosine part of the function $y = 3/2 + 3\cos(2\pi\,1\,t) + 4\sin(2\pi\,3\,t)$.

(6) that $A[3] = 7.54018*10^{\wedge}-14-7.97885\ I$ is purely imaginary as it corresponds to the sine part of the function $y = 3/2 + 3\cos(2\pi\,1\,t) + 4\sin(2\pi\,3\,t)$.

(7) that $A[3] = 7.54018*10^{\wedge}-14-7.97885$ I and $A[7] = -2.69895*10^{\wedge}-13+7.97885$ I are complex conjugate of each other, as expected from Eq. (1.10): $A(n_{max} - n) = A^*(n)$.

(8) that we need not use true periodicities of the function $y = 3/2 + 3 \cos(2\pi\ 1\ t) + 4 \sin(2\pi\ 3\ t)$ in calculating the frequency content of the data.

(9) that absolute values of $A[1]$ and $A[3]$ are Abs $[A[1]] = 5.98413$ and Abs $[A[3]] = 7.97885$. These 2 numbers give us an idea of relative dominance of the 2 frequencies in the oscillatory data. These absolute values are proportional to, but *not* equal to 3 and 4 respectively as one may anticipate from $y = 3/2 + \mathbf{3} \cos(2\pi\ 1\ t) + \mathbf{4} \sin(2\pi\ 3\ t)$.

1.6   Discrete Fourier Transform: Example V

In this section, we take up the function $y = \cos(4(t - 3))\ e^{-0.5(t-3)^2}$ of time t. Using Program number 1.5, we first plot the function; see Fig. 1.5. We then turn the analytic function $y(t)$ into numerical data denoted by $y[i]$ for a discrete set of values of time t in the interval 0–5 s. These discrete data are then fed to the rest of the program that calculates (the complex numbers) $A[n]$'s, and absolute values of them which constitute what is called power spectrum, see Table 1.5. We find that frequencies that are present are: $n = 0$, 0.4, 0.6, 0.8 and 1.0 Hz. Also present are non-zero values of $A[n]$ for unphysical values of n, which are 19, 19.2, 19.4, 19.6 and 20 Hz. See Fig. 1.6.

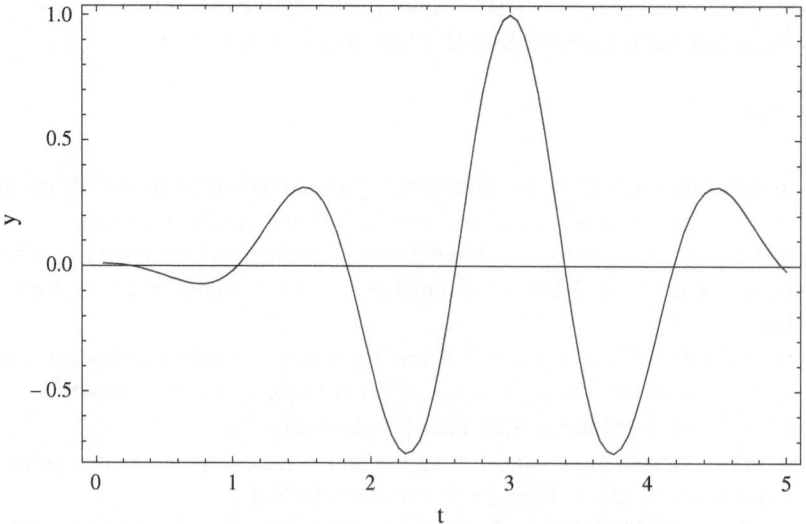

Fig. 1.5 Showing the function $y = \cos(4(t - 3))\ e^{-0.5(t-3)^2}$ as function of time t

Table 1.5 Discrete Fourier transform for the numerical data for $y = \cos(4(t - 3))$ $e^{-0.5(t-3)^2}$

n	$A[n]$	$Abs\,[A[n]]$
0.0	**0.0107429**	**0.0107**
0.2	0.000933615+0.00356764 I	0.0037
0.4	**0.060205-0.164987 I**	**0.1756**
0.6	**0.155327+0.451984 I**	**0.4779**
0.8	**-0.239247-0.184375 I**	**0.3020**
1.0	**0.0337896-0.00776122 I**	**0.0347**
1.2	-0.00412323-0.00443572 I	0.0061
1.4	-0.00284703-0.00335537 I	0.0044
1.6	-0.00233027-0.00242338 I	0.0034
1.8	-0.00192743-0.00186544 I	0.0027
2.0	-0.00162705-0.00150311 I	0.0022
2.2	-0.0014016-0.00125233 I	0.0019
2.4	-0.00122945-0.00106969 I	0.0016
2.6	-0.00109552-0.000931142 I	0.0014
2.8	-0.000989459-0.000822521 I	0.0013
3.0	-0.00090414-0.000735013 I	0.0012
3.2	-0.000834535-0.000662897 I	0.0011
3.4	-0.000777042-0.000602314 I	0.0010
3.6	-0.00072903-0.00055058 I	0.0009
3.8	-0.000688544-0.000505773 I	0.0009
4.0	-0.000654109-0.000466483 I	0.0008
4.2	-0.000624595-0.000431657 I	0.0008
4.4	-0.000599123-0.000400489 I	0.0007
4.6	-0.000577007-0.000372355 I	0.0007
4.8	-0.000557698-0.000346763 I	0.0007
5.0	-0.000540759-0.000323321 I	0.0006
5.2	-0.000525834-0.000301712 I	0.0006
5.4	-0.000512634-0.000281677 I	0.0006
5.6	-0.00050092-0.000263001 I	0.0006
5.8	-0.000490496-0.000245506 I	0.0005
6.0	-0.000481197-0.000229043 I	0.0005
6.2	-0.000472886-0.000213483 I	0.0005
6.4	-0.000465446-0.000198718 I	0.0005
6.6	-0.000458779-0.000184655 I	0.0005
6.8	-0.000452803-0.000171212 I	0.0005

(continued)

Table 1.5 (continued)

n	$A[n]$	$Abs\,[A[n]]$
7.0	-0.000447446-0.000158317 I	0.0005
7.2	-0.000442647-0.000145909 I	0.0005
7.4	-0.000438354-0.000133931 I	0.0005
7.6	-0.000434522-0.000122336 I	0.0005
7.8	-0.000431113-0.000111077 I	0.0004
8.0	-0.000428094-0.000100116 I	0.0004
8.2	-0.000425437-0.0000894158 I	0.0004
8.4	-0.000423117-0.0000789427 I	0.0004
8.6	-0.000421114-0.0000686661 I	0.0004
8.8	-0.00041941-0.0000585571 I	0.0004
9.0	-0.000417992-0.0000485889 I	0.0004
9.2	-0.000416847-0.0000387358 I	0.0004
9.4	-0.000415966-0.0000289735 I	0.0004
9.6	-0.000415341-0.0000192786 I	0.0004
9.8	-0.000414968-9.62822*10^-6 I	0.0004
10.0	-0.000414845-7.01959*10^-16 I	0.0004
10.2	-0.000414968+9.62822*10^-6 I	0.0004
10.4	-0.000415341+0.0000192786 I	0.0004
10.6	-0.000415966+0.0000289735 I	0.0004
10.8	-0.000416847+0.0000387358 I	0.0004
11.0	-0.000417992+0.0000485889 I	0.0004
11.2	-0.00041941+0.0000585571 I	0.0004
11.4	-0.000421114+0.0000686661 I	0.0004
11.6	-0.000423117+0.0000789427 I	0.0004
11.8	-0.000425437+0.0000894158 I	0.0004
12.0	-0.000428094+0.000100116 I	0.0004
12.2	-0.000431113+0.000111077 I	0.0004
12.4	-0.000434522+0.000122336 I	0.0005
12.6	-0.000438354+0.000133931 I	0.0005
12.8	-0.000442647+0.000145909 I	0.0005
13.0	-0.000447446+0.000158317 I	0.0005
13.2	-0.000452803+0.000171212 I	0.0005
13.4	-0.000458779+0.000184655 I	0.0005
13.6	-0.000465446+0.000198718 I	0.0005
13.8	-0.000472886+0.000213483 I	0.0005

(continued)

Table 1.5 (continued)

n	A[n]	Abs [A[n]]
14.0	-0.000481197+0.000229043 I	0.0005
14.2	-0.000490496+0.000245506 I	0.0005
14.4	-0.00050092+0.000263001 I	0.0006
14.6	-0.000512634+0.000281677 I	0.0006
14.8	-0.000525834+0.000301712 I	0.0006
15.0	-0.000540759+0.000323321 I	0.0006
15.2	-0.000557698+0.000346763 I	0.0007
15.4	-0.000577007+0.000372355 I	0.0007
15.6	-0.000599123+0.000400489 I	0.0007
15.8	-0.000624595+0.000431657 I	0.0008
16.0	-0.000654109+0.000466483 I	0.0008
16.2	-0.000688544+0.000505773 I	0.0009
16.4	-0.00072903+0.00055058 I	0.0009
16.6	-0.000777042+0.000602314 I	0.0010
16.8	-0.000834535+0.000662897 I	0.0011
17.0	-0.00090414+0.000735013 I	0.0012
17.2	-0.000989459+0.000822521 I	0.0013
17.4	-0.00109552+0.000931142 I	0.0014
17.6	-0.00122945+0.00106969 I	0.0016
17.8	-0.0014016+0.00125233 I	0.0019
18.0	-0.00162705+0.00150311 I	0.0022
18.2	-0.00192743+0.00186544 I	0.0027
18.4	-0.00233027+0.00242338 I	0.0034
18.6	-0.00284703+0.00335537 I	0.0044
18.8	-0.00412323+0.00443572 I	0.0061
19.0	**0.0337896+0.00776122 I**	**0.0347**
19.2	**-0.239247+0.184375 I**	**0.3020**
19.4	**0.155327-0.451984 I**	**0.4779**
19.6	**0.060205+0.164987 I**	**0.1756**
19.8	0.000933615-0.00356764 I	0.0037
20.0	**0.0107429+1.16952*10^-15 I**	**0.0107**

Key results are in bold face. $I = \sqrt{(-1)}$.

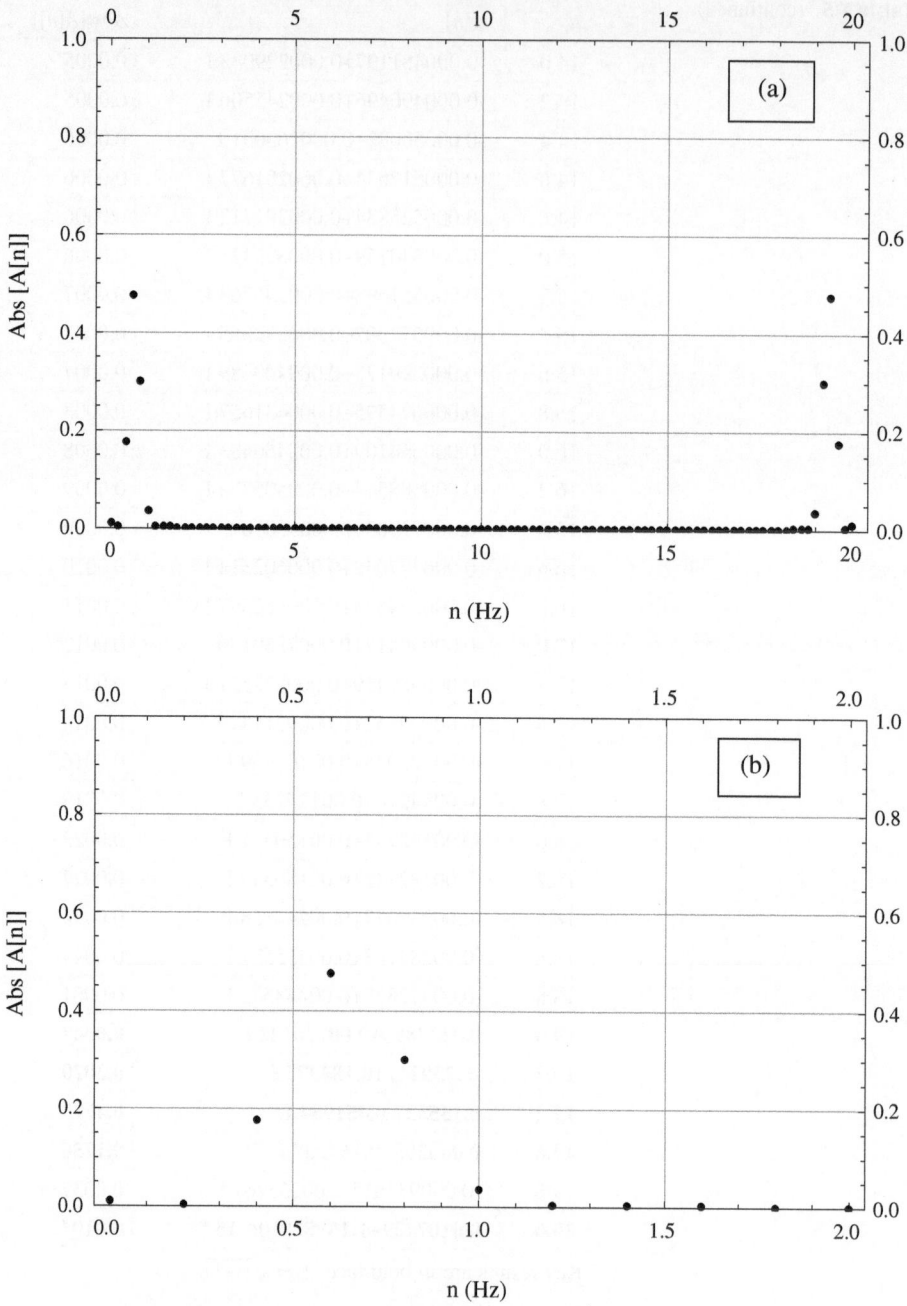

Fig. 1.6 Discrete Fourier transform for the numerical data for $y = \cos(4(t-3))\, e^{-0.5(t-3)^2}$

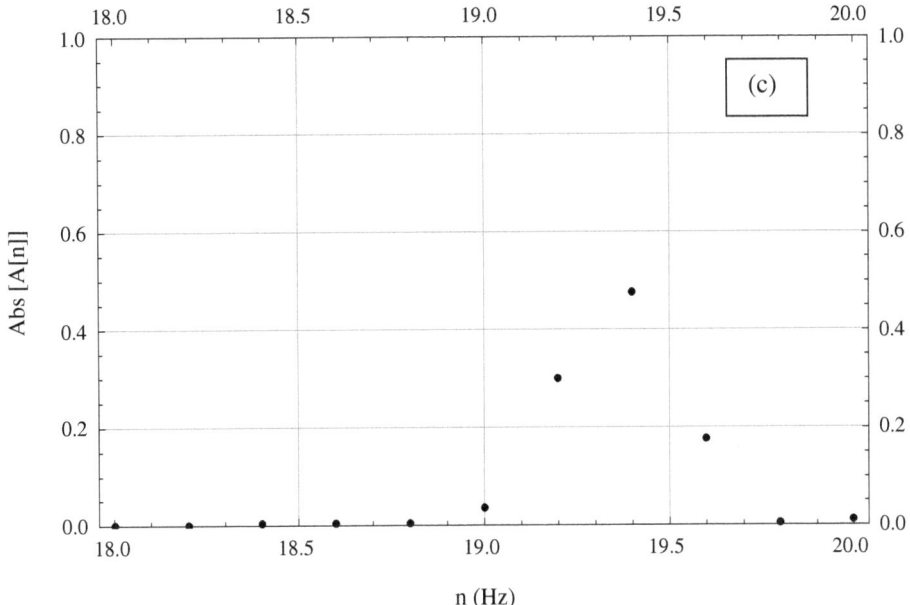

Fig. 1.6   (continued)

Program number 1.5

```
h=N[5/100];
i=0;
Table[{i=i+1,t=t+h,y[i]=(Cos[4*(t-3)])*Exp[-0.5*(t-3)^2]},{t,0,5-h,h}];
TableForm[%,TableSpacing->{3,3},
TableHeadings->{None,{"i","t","y[i]"}}]
```

```
i=0;
ListLinePlot[Table[{i=i+1;t=t+h,y[i]},{t,0,5-h,h}],
Frame->True,FrameLabel->{"t","y"}]
```

```
Z=N[Exp[-2*Pi*I/100]];
```

```
Table[{n1=n1+1;n=N[n1/5],A[n1]=h*(1/Sqrt[2*Pi])*(Sum[y[k]*(Z^(n1*k)),
{k,1,100,1}]),Ab[n1]=Abs[A[n1]]},{n1,0-1,100-1,1}];
TableForm[%,TableSpacing->{3,3},
```

TableHeadings->{None,{"n","A[n]","Abs [A[n]]"}}]

ListPlot[Table[{n1=n1+1;n=n1/5,Ab[n1]},{n1,0-1,100-1,1}],
Frame->True,FrameLabel->{"n (Hz)","Abs [A[n]"},PlotStyle->{Black},
GridLines->Automatic,FrameTicks->All,PlotRange->{0,1}]

ListPlot[Table[{n1=n1+1;n=n1/5,Ab[n1]},{n1,0-1,10-1,1}],
Frame->True,FrameLabel->{"n (Hz)","Abs [A[n]"},PlotStyle->{Black},
GridLines->Automatic,FrameTicks->All,PlotRange->{0,1}]

ListPlot[Table[{n1=n1+1;n=n1/5,Ab[n1]},{n1,90-1,100-1,1}],
Frame->True,FrameLabel->{"n (Hz)","Abs [A[n]"},PlotStyle->{Black},
GridLines->Automatic,FrameTicks->All,PlotRange->{0,1}]

1.7   Fourier Series Approximation

In this section, we use Program number 1.6. We first turn the function $y = \cos(4(t - 3))\, e^{-0.5(t-3)^2}$ to a data set for time t in the interval 0–5 s and plot the data as in Fig. 1.7. We also plot the function as in Fig. 1.8 to reveal that it is a localized function, and is *not* a periodic function of time t. We make use of the (physical rather than unphysical) frequencies that are present in the data for the function $y = \cos(4(t-3))\, e^{-0.5(t-3)^2}$. The frequencies are: $n = 0$, 0.4, 0.6, 0.8 and 1.0 Hz. And we obtain corresponding Fourier coefficients a_0, a_2, a_3, a_4, a_5, b_2, b_3, b_4 and b_5. The coefficients turn out to be $a_0 = 0.0111038$, $a_2 = 0.0606972$, $a_3 = 0.156071$, $a_4 = -0.239548$, $a_5 = 0.0342126$, $b_2 = 0.165431$, $b_3 = -0.453173$, $b_4 = 0.184877$, $b_5 = 0.00779804$. In evaluating these coefficients, we use standard expressions for evaluating Fourier coefficients, and Simpson's rule for numerical integration.

Simpson's rule for numerical integration is as follows. If v_0, v_1, v_2, v_3, ..., v_{n-2}, v_{n-1}, v_n are values of v for $x = x_0$, $x_0 + h$, $x_0 + 2h$, $x_0 + 3h$, ..., $x_0 + (n-2)h$, $x_0 + (n-1)h$, $x_n = x_0 + nh$ respectively, Simpson's rule for numerical integration gives

$$\int_{x_0}^{x_n} v\, dx = \frac{h}{3}\left[v_0 + 4(v_1 + v_3 + v_5 + \cdots + v_{n-1}) + 2(v_2 + v_4 + v_6 + \cdots + v_{n-2}) + v_n\right]$$

which is well known to yield excellent agreement with standard result where available.

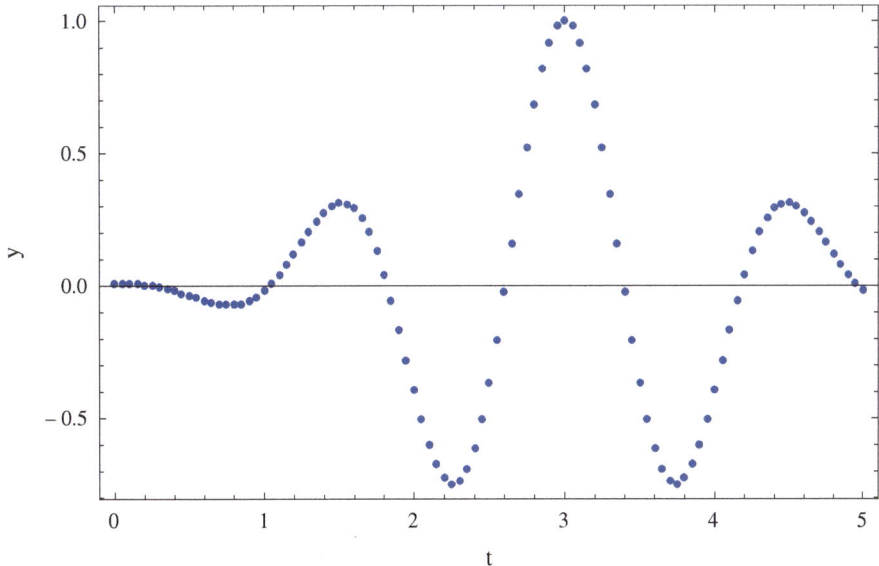

Fig. 1.7 Showing discrete set of data for the function $y = \cos(4(t-3)) \, e^{-0.5(t-3)^2}$ in the time interval $t = 0\text{–}5$ s. Using plot command p1 of Program number 1.6

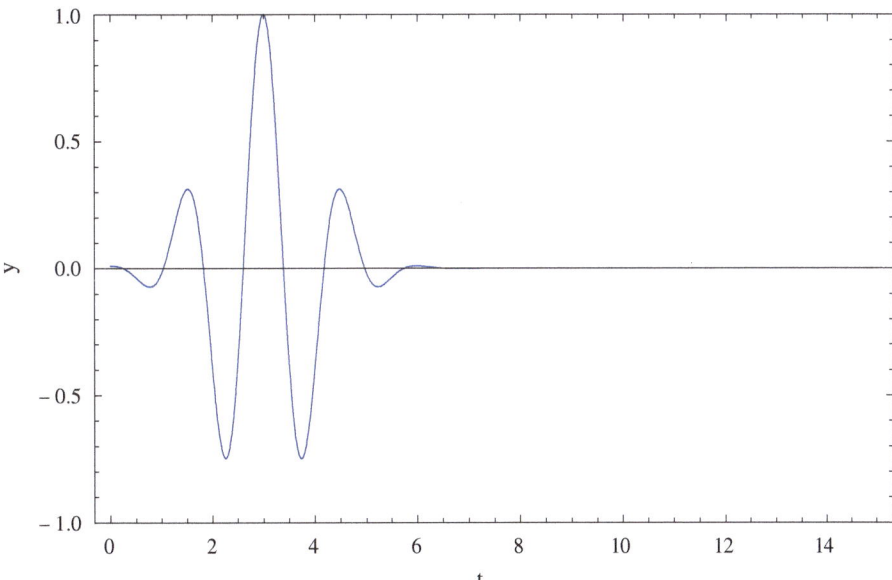

Fig. 1.8 Showing that the function $y = \cos(4(t-3)) \, e^{-0.5(t-3)^2}$ is localized and is *not* a periodic function of time t. Using plot command p2 of Program number 1.6

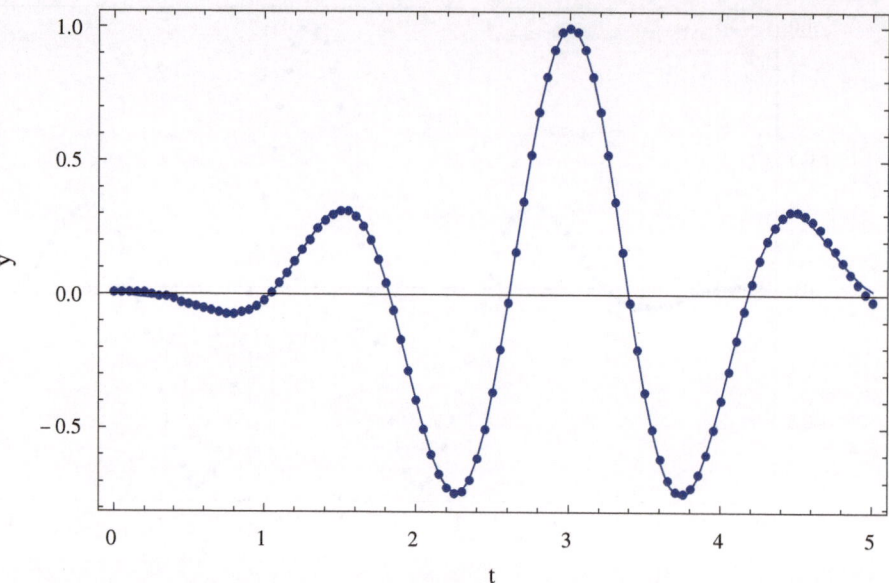

Fig. 1.9 Dotted are data for the function $y = \cos(4(t-3))\, e^{-0.5(t-3)^2}$, shown in Fig. 1.7, obtained using plot command p1 of Program number 1.6. The **curve** shows Fourier series approximation obtained using plot command p3 of Program number 1.6

Fourier series approximation is as follows. If $f(t)$ is a function of time t, we can approximate $f(t)$ in the time interval 0 to T by the following series:

$$f(t) = \frac{a_0}{2} + \sum_{n=1}^{\infty} a_n \cos\left(\frac{2\pi nt}{T}\right) + \sum_{n=1}^{\infty} b_n \sin\left(\frac{2\pi nt}{T}\right)$$

where a_0, a_n's and b_n's are given by (Fig. 1.9)

$$a_0 = \frac{2}{T}\int_0^T f(t)\,dt$$

$$a_n = \frac{2}{T}\int_0^T f(t)\cos\left(\frac{2\pi nt}{T}\right)dt$$

$$b_n = \frac{2}{T}\int_0^T f(t)\sin\left(\frac{2\pi nt}{T}\right)dt$$

Program number 1.6

```
h=N[5/100];
i=-1;
Table[{i=i+1,t=t+h,y[i]=(Cos[4*(t-3)])*Exp[-0.5*(t-3)^2],
sin2t[i]=Sin[2*Pi*2*t/5];cos2t[i]=Cos[2*Pi*2*t/5];
sin3t[i]=Sin[2*Pi*3*t/5];cos3t[i]=Cos[2*Pi*3*t/5];
sin4t[i]=Sin[2*Pi*4*t/5];cos4t[i]=Cos[2*Pi*4*t/5];
sin5t[i]=Sin[2*Pi*5*t/5];cos5t[i]=Cos[2*Pi*5*t/5];},{t,0-h,5-h,h}];
TableForm[%,TableHeadings->{None,{"i","t","y[i]"}}]


i=-1;
p1=ListPlot[Table[{i=i+1;t=t+h,y[i]},{t,0-h,5-h,h}],
Frame->True,FrameLabel->{"t","y"}]
p2=Plot[(Cos[4*(t-3)])*Exp[-0.5*(t-3)^2],{t,0,15},
PlotRange->{-1,1},Frame->True,FrameLabel->{"t","y"}]


a0=(h/(3*2.5))*(y[0]+4*(Sum[y[2*i+1],{i,0,49}])+2*(Sum[y[2*i],{i,1,49}])+y[100])


a2=(h/(3*2.5))*(y[0]*cos2t[0]+4*(Sum[y[2*i+1]*cos2t[2*i+1],{i,0,49}])+
2*(Sum[y[2*i]*cos2t[2*i],{i,1,49}])+y[100]*cos2t[100])


a3=(h/(3*2.5))*(y[0]*cos3t[0]+4*(Sum[y[2*i+1]*cos3t[2*i+1],{i,0,49}])+
2*(Sum[y[2*i]*cos3t[2*i],{i,1,49}])+y[100]*cos3t[100])


a4=(h/(3*2.5))*(y[0]*cos4t[0]+4*(Sum[y[2*i+1]*cos4t[2*i+1],{i,0,49}])+
2*(Sum[y[2*i]*cos4t[2*i],{i,1,49}])+y[100]*cos4t[100])


a5=(h/(3*2.5))*(y[0]*cos5t[0]+4*(Sum[y[2*i+1]*cos5t[2*i+1],{i,0,49}])+
2*(Sum[y[2*i]*cos5t[2*i],{i,1,49}])+y[100]*cos5t[100])


b2=(h/(3*2.5))*(y[0]*sin2t[0]+4*(Sum[y[2*i+1]*sin2t[2*i+1],{i,0,49}])+
```

```
2*(Sum[y[2*i]*sin2t[2*i],{i,1,49}])+y[100]*sin2t[100])


b3=(h/(3*2.5))*(y[0]*sin3t[0]+4*(Sum[y[2*i+1]*sin3t[2*i+1],{i,0,49}])+
2*(Sum[y[2*i]*sin3t[2*i],{i,1,49}])+y[100]*sin3t[100])


b4=(h/(3*2.5))*(y[0]*sin4t[0]+4*(Sum[y[2*i+1]*sin4t[2*i+1],{i,0,49}])+
2*(Sum[y[2*i]*sin4t[2*i],{i,1,49}])+y[100]*sin4t[100])


b5=(h/(3*2.5))*(y[0]*sin5t[0]+4*(Sum[y[2*i+1]*sin5t[2*i+1],{i,0,49}])+
2*(Sum[y[2*i]*sin5t[2*i],{i,1,49}])+y[100]*sin5t[100])


p3=ListLinePlot[Table[{t=t+h,
a0/2+a2*Cos[2*Pi*2*t/5]+a3*Cos[2*Pi*3*t/5]+
a4*Cos[2*Pi*4*t/5]+a5*Cos[2*Pi*5*t/5]+
b2*Sin[2*Pi*2*t/5]+b3*Sin[2*Pi*3*t/5]+
b4*Sin[2*Pi*4*t/5]+b5*Sin[2*Pi*5*t/5]},{t,0-h,5-h,h}],
Frame->True,FrameLabel->{"t","y"}];
Show[p1,p3]


p4=ListLinePlot[Table[{t=t+h,
a0/2+a2*Cos[2*Pi*2*t/5]+a3*Cos[2*Pi*3*t/5]+
a4*Cos[2*Pi*4*t/5]+a5*Cos[2*Pi*5*t/5]+
b2*Sin[2*Pi*2*t/5]+b3*Sin[2*Pi*3*t/5]+
b4*Sin[2*Pi*4*t/5]+b5*Sin[2*Pi*5*t/5]},{t,0-h,15-h,h}],
Frame->True,FrameLabel->{"t","y"}]
```

To obtain Fourier series approximation, we first need to know the frequencies that we need. Here lies a need of Fourier transform. Now that we have Fourier series for the function $y = \cos(4(t-3))\, e^{-0.5(t-3)^2}$, we can plot the Fourier series in extended interval of time t and obtain what is shown in Fig. 1.10.

We find that a function need not be periodic to be expressed analytically as a Fourier series. But after expressing the function as a Fourier series, we can plot it in extended interval of time t and get repeated or periodic plot of the original non-periodic function.

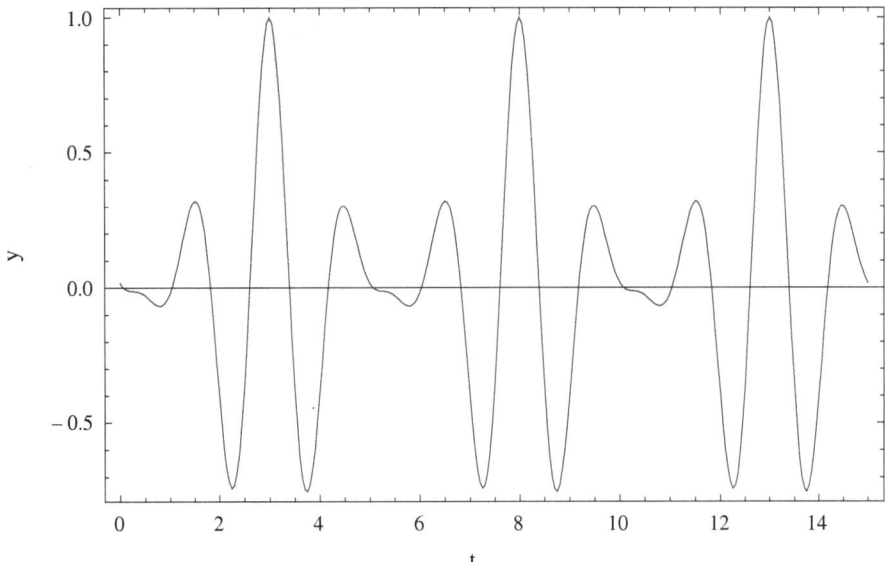

Fig. 1.10 Fourier series for the function $y = \cos(4(t-3))\, e^{-0.5(t-3)^2}$, plotted in extended interval of time t, obtained using plot command p4 of Program number 1.6

Motion and Power Spectrum of Driven Damped Oscillator: Analytical and Numerical Account

Abstract

This chapter contains analytical and numerical solutions of differential equation of motion of driven damped oscillator using 4th order Runge-Kutta method. Data of the numerical solution are fed to a discrete Fourier transform program to obtain frequency content of the system. Programs were written in Mathematica to achieve these.

2.1 Analytical Solution of Differential Equation of Motion of Driven Damped Oscillator

For damped harmonic oscillator without the driven motion, differential equation of motion is

$$m\frac{d^2y}{dt^2} + B\frac{dy}{dt} + k\,y = 0 \tag{2.1}$$

where m is mass of the particle executing motion, y is displacement of the particle from equilibrium, t is time, B is damping coefficient of the velocity dependent damping force $-B\frac{dy}{dt}$, k is constant of Hooke's law force $-\,k\,y$.

Analytic solution for displacement y as a function of time for Eq. (2.1) is

$$y = y_{\max}e^{-Bt/(2m)}\cos(\omega_d\,t) \tag{2.2}$$

where $\omega_d = \sqrt{\frac{k}{m} - \left(\frac{B}{2m}\right)^2}$ is angular frequency of damped harmonic motion. $\omega_0 = \sqrt{\frac{k}{m}}$ is angular frequency of simple harmonic motion in absence of the damping force.

If a periodic driving force $F_{\max}\cos(\omega\,t)$ is introduced to the damped harmonic motion, the differential equation of motion becomes

© The Author(s), under exclusive license to Springer Nature Switzerland AG 2024
S. Chowdhury and A. Al Sakib, *Numerical Exploration of Fourier Transform and Fourier Series*, Synthesis Lectures on Mathematics & Statistics,
https://doi.org/10.1007/978-3-031-34664-4_2

$$m\frac{d^2y}{dt^2} + B\frac{dy}{dt} + k\,y = F_{max}\cos(\omega\,t) \tag{2.3}$$

Analytic solution for displacement y as a function of time t for Eq. (2.3) is

$$y = \frac{F_{max}\sin(\omega\,t - \phi)}{\sqrt{m^2(\omega^2 - \omega_0^2)^2 + B^2\omega^2}} = \frac{F_{max}\sin(\omega\,t - \phi)}{G} \tag{2.4}$$

where

$$\phi = \tan^{-1}\left(\frac{\sqrt{G^2 - (B\omega)^2}}{B\omega}\right) = \tan^{-1}\left(\frac{\omega^2 - \omega_0^2}{b\omega}\right) \tag{2.5}$$

Here $b = B/m$. Equations (2.1)–(2.5) have been adapted from Ref. [2].

2.2 Numerical Solution of Differential Equation of Motion of the Driven Damped Oscillator

We now use 4th order Runge–Kutta method in numerically solving Eq. (2.3) written as

$$\frac{d^2y}{dt^2} + b\frac{dy}{dt} + \omega_0^2 y = a\cos(\omega\,t) \tag{2.6}$$

To this end, we now rewrite Eq. (2.6) as 2 1st order differential equations as

$$\frac{dv}{dt} = -\omega_0^2 y - bv + a\cos(\omega\,t) \tag{2.7}$$

and

$$\frac{dy}{dt} = v \tag{2.8}$$

We now gather that iteration for numerical solution of $\frac{dy}{dx} = H(x, y)$ is given by

$$y_{n+1} = y_n + (k_1 + 2k_2 + 2k_3 + k_4)/6 \tag{2.9}$$

where

$$k_1 = h\,H(x_n, y_n) \tag{2.10}$$

$$k_2 = h\,H(x_n + h/2, y_n + k_1/2) \tag{2.11}$$

$$k_3 = h\,H(x_n + h/2, y_n + k_2/2) \tag{2.12}$$

$$k_4 = h\, H(x_n + h,\, y_n + k_3) \qquad\qquad (2.13)$$

As such, we can solve Eq. (2.7) for v taking

$$H = -\omega_0^2 y - bv + a\cos(\omega\, t)$$

for which

$$k_1 = h\!\left(-\omega_0^2 y - bv + a\, \cos(\omega\, t)\right)$$

$$k_2 = h\!\left(-\omega_0^2 y - b(v + k_1/2) + a\cos(\omega(t + h/2))\right)$$

$$k_3 = h\!\left(-\omega_0^2 y - b(v + k_2/2) + a\cos(\omega(t + h/2))\right)$$

$$k_4 = h\!\left(-\omega_0^2 y - b(v + k_3) + a\cos(\omega(t + h))\right)$$

As such, in Program number 2.1, we write the above 4 k's as

$$k1v=h^*\!\left((-w0\char`^2)^*y-b^*v+a^*Cos\!\left[w^*(t\text{-}h)\right]\right)$$

$$k2v=h^*\!\left((-w0\char`^2)^*y-b^*(v+k1v/2)+a^*Cos\!\left[w^*(t\text{-}h+h/2)\right]\right)$$

$$k3v=h^*\!\left((-w0\char`^2)^*y-b^*(v+k2v/2)+a^*Cos\!\left[w^*(t\text{-}h+h/2)\right]\right)$$

$$k4v=h^*\!\left((-w0\char`^2)^*y-b^*(v+k3v)+a^*Cos\!\left[w^*(t\text{-}h+h)\right]\right)$$

to get

$$v=v+(k1v+2*k2v+2*k3v+k4v)/6$$

And we can solve Eq. (2.8) for y taking $H = v$ for which we use

$$k_1 = k_2 = k_3 = k_4 = h\, v$$

For further studies on 4th order Runge–Kutta method, see for example Ref. [3].

2.3 Motion and Power Spectrum of Driven Damped Oscillator: Typical Features

Tables 2.1, 2.2 and Figs. 2.1, 2.2 show *typical* data and *typical* plots for displacement of driven damped oscillator as function of time and corresponding power spectrum showing frequency content of the system. To obtain these, we have written Program number 2.1.

Table 2.1 *Typical* data table obtained using Program number 2.1 giving values of *y* as function of time *t* numerically obtained using 4th order Runge–Kutta method

i	t	y	i	t	y
1	0.05	0.9110	201	10.05	0.0949
2	0.10	0.7459	202	10.10	0.0917
3	0.15	0.5239	203	10.15	0.0868
4	0.20	0.2690	204	10.20	0.0804
5	0.25	0.0067	205	10.25	0.0728
6	0.30	-0.2377	206	10.30	0.0641
7	0.35	-0.4424	207	10.35	0.0547
8	0.40	-0.5902	208	10.40	0.0447
9	0.45	-0.6704	209	10.45	0.0345
10	0.50	-0.6795	210	10.50	0.0241
11	0.55	-0.6210	211	10.55	0.0138
12	0.60	-0.5046	212	10.60	0.0035
13	0.65	-0.3453	213	10.65	-0.0066
14	0.70	-0.1614	214	10.70	-0.0165
15	0.75	0.0272	215	10.75	-0.0262
16	0.80	0.2013	216	10.80	-0.0355
17	0.85	0.3440	217	10.85	-0.0445
18	0.90	0.4423	218	10.90	-0.0531
19	0.95	0.4882	219	10.95	-0.0612
20	1.00	0.4793	220	11.00	-0.0686
21	1.05	0.4185	221	11.05	-0.0752
22	1.10	0.3137	222	11.10	-0.0809
23	1.15	0.1769	223	11.15	-0.0854
24	1.20	0.0226	224	11.20	-0.0887

(continued)

Table 2.1 (continued)

i	t	y	i	t	y
25	1.25	-0.1337	225	11.25	-0.0906
26	1.30	-0.2768	226	11.30	-0.0910
27	1.35	-0.3935	227	11.35	-0.0898
28	1.40	-0.4735	228	11.40	-0.0869
29	1.45	-0.5106	229	11.45	-0.0824
30	1.50	-0.5030	230	11.50	-0.0764
31	1.55	-0.4529	231	11.55	-0.0690
32	1.60	-0.3667	232	11.60	-0.0602
33	1.65	-0.2537	233	11.65	-0.0504
34	1.70	-0.1253	234	11.70	-0.0398
35	1.75	0.0063	235	11.75	-0.0285
36	1.80	0.1290	236	11.80	-0.0168
37	1.85	0.2323	237	11.85	-0.0050
38	1.90	0.3083	238	11.90	0.0067
39	1.95	0.3518	239	11.95	0.0181
40	2.00	0.3612	240	12.00	0.0290
41	2.05	0.3380	241	12.05	0.0394
42	2.10	0.2869	242	12.10	0.0489
43	2.15	0.2150	243	12.15	0.0576
44	2.20	0.1309	244	12.20	0.0653
45	2.25	0.0440	245	12.25	0.0720
46	2.30	-0.0368	246	12.30	0.0776
47	2.35	-0.1035	247	12.35	0.0820
48	2.40	-0.1500	248	12.40	0.0854
49	2.45	-0.1726	249	12.45	0.0875
50	2.50	-0.1702	250	12.50	0.0884
51	2.55	-0.1442	251	12.55	0.0880
52	2.60	-0.0985	252	12.60	0.0864
53	2.65	-0.0385	253	12.65	0.0835
54	2.70	0.0288	254	12.70	0.0793
55	2.75	0.0963	255	12.75	0.0738
56	2.80	0.1568	256	12.80	0.0671
57	2.85	0.2043	257	12.85	0.0593
58	2.90	0.2341	258	12.90	0.0505

(continued)

Table 2.1 (continued)

i	t	y	i	t	y
59	2.95	0.2434	259	12.95	0.0407
60	3.00	0.2316	260	13.00	0.0302
61	3.05	0.1998	261	13.05	0.0191
62	3.10	0.1512	262	13.10	0.0077
63	3.15	0.0905	263	13.15	-0.0040
64	3.20	0.0231	264	13.20	-0.0156
65	3.25	-0.0451	265	13.25	-0.0270
66	3.30	-0.1083	266	13.30	-0.0379
67	3.35	-0.1615	267	13.35	-0.0482
68	3.40	-0.2009	268	13.40	-0.0576
69	3.45	-0.2240	269	13.45	-0.0660
70	3.50	-0.2301	270	13.50	-0.0733
71	3.55	-0.2199	271	13.55	-0.0794
72	3.60	-0.1956	272	13.60	-0.0841
73	3.65	-0.1606	273	13.65	-0.0875
74	3.70	-0.1191	274	13.70	-0.0894
75	3.75	-0.0754	275	13.75	-0.0900
76	3.80	-0.0339	276	13.80	-0.0891
77	3.85	0.0016	277	13.85	-0.0869
78	3.90	0.0284	278	13.90	-0.0834
79	3.95	0.0444	279	13.95	-0.0787
80	4.00	0.0493	280	14.00	-0.0727
81	4.05	0.0435	281	14.05	-0.0657
82	4.10	0.0289	282	14.10	-0.0577
83	4.15	0.0078	283	14.15	-0.0489
84	4.20	-0.0165	284	14.20	-0.0392
85	4.25	-0.0407	285	14.25	-0.0290
86	4.30	-0.0616	286	14.30	-0.0183
87	4.35	-0.0765	287	14.35	-0.0072
88	4.40	-0.0833	288	14.40	0.0040
89	4.45	-0.0808	289	14.45	0.0152
90	4.50	-0.0688	290	14.50	0.0262
91	4.55	-0.0481	291	14.55	0.0368
92	4.60	-0.0203	292	14.60	0.0469

(continued)

Table 2.1 (continued)

i	t	y	i	t	y
93	4.65	0.0124	293	14.65	0.0563
94	4.70	0.0474	294	14.70	0.0649
95	4.75	0.0817	295	14.75	0.0724
96	4.80	0.1127	296	14.80	0.0787
97	4.85	0.1379	297	14.85	0.0838
98	4.90	0.1555	298	14.90	0.0875
99	4.95	0.1646	299	14.95	0.0898
100	5.00	0.1646	300	15.00	0.0906
101	5.05	0.1561	301	15.05	0.0900
102	5.10	0.1401	302	15.10	0.0879
103	5.15	0.1184	303	15.15	0.0844
104	5.20	0.0930	304	15.20	0.0796
105	5.25	0.0660	305	15.25	0.0735
106	5.30	0.0396	306	15.30	0.0662
107	5.35	0.0157	307	15.35	0.0580
108	5.40	-0.0042	308	15.40	0.0488
109	5.45	-0.0193	309	15.45	0.0390
110	5.50	-0.0291	310	15.50	0.0285
111	5.55	-0.0338	311	15.55	0.0176
112	5.60	-0.0341	312	15.60	0.0065
113	5.65	-0.0311	313	15.65	-0.0047
114	5.70	-0.0261	314	15.70	-0.0159
115	5.75	-0.0208	315	15.75	-0.0268
116	5.80	-0.0164	316	15.80	-0.0372
117	5.85	-0.0141	317	15.85	-0.0471
118	5.90	-0.0149	318	15.90	-0.0563
119	5.95	-0.0192	319	15.95	-0.0646
120	6.00	-0.0271	320	16.00	-0.0720
121	6.05	-0.0382	321	16.05	-0.0782
122	6.10	-0.0517	322	16.10	-0.0832
123	6.15	-0.0665	323	16.15	-0.0869

(continued)

Table 2.1 (continued)

i	t	y	i	t	y
124	6.20	-0.0814	324	16.20	-0.0892
125	6.25	-0.0950	325	16.25	-0.0902
126	6.30	-0.1060	326	16.30	-0.0897
127	6.35	-0.1136	327	16.35	-0.0877
128	6.40	-0.1167	328	16.40	-0.0844
129	6.45	-0.1151	329	16.45	-0.0797
130	6.50	-0.1087	330	16.50	-0.0738
131	6.55	-0.0977	331	16.55	-0.0666
132	6.60	-0.0829	332	16.60	-0.0584
133	6.65	-0.0651	333	16.65	-0.0493
134	6.70	-0.0454	334	16.70	-0.0394
135	6.75	-0.0250	335	16.75	-0.0288
136	6.80	-0.0050	336	16.80	-0.0179
137	6.85	0.0136	337	16.85	-0.0066
138	6.90	0.0300	338	16.90	0.0047
139	6.95	0.0436	339	16.95	0.0160
140	7.00	0.0542	340	17.00	0.0270
141	7.05	0.0617	341	17.05	0.0375
142	7.10	0.0665	342	17.10	0.0475
143	7.15	0.0689	343	17.15	0.0567
144	7.20	0.0696	344	17.20	0.0650
145	7.25	0.0692	345	17.25	0.0722
146	7.30	0.0683	346	17.30	0.0784
147	7.35	0.0675	347	17.35	0.0833
148	7.40	0.0672	348	17.40	0.0869
149	7.45	0.0676	349	17.45	0.0891
150	7.50	0.0688	350	17.50	0.0900
151	7.55	0.0706	351	17.55	0.0894
152	7.60	0.0727	352	17.60	0.0875
153	7.65	0.0747	353	17.65	0.0841
154	7.70	0.0759	354	17.70	0.0795
155	7.75	0.0760	355	17.75	0.0736
156	7.80	0.0744	356	17.80	0.0665

(continued)

Table 2.1 (continued)

i	t	y	i	t	y
157	7.85	0.0707	357	17.85	0.0584
158	7.90	0.0647	358	17.90	0.0493
159	7.95	0.0562	359	17.95	0.0395
160	8.00	0.0455	360	18.00	0.0290
161	8.05	0.0327	361	18.05	0.0181
162	8.10	0.0183	362	18.10	0.0068
163	8.15	0.0028	363	18.15	-0.0045
164	8.20	-0.0131	364	18.20	-0.0158
165	8.25	-0.0288	365	18.25	-0.0268
166	8.30	-0.0436	366	18.30	-0.0374
167	8.35	-0.0571	367	18.35	-0.0474
168	8.40	-0.0687	368	18.40	-0.0567
169	8.45	-0.0781	369	18.45	-0.0650
170	8.50	-0.0852	370	18.50	-0.0724
171	8.55	-0.0899	371	18.55	−0.0785
172	8.60	-0.0923	372	18.60	-0.0835
173	8.65	-0.0927	373	18.65	-0.0871
174	8.70	-0.0914	374	18.70	-0.0893
175	8.75	-0.0886	375	18.75	-0.0901
176	8.80	-0.0847	376	18.80	-0.0895
177	8.85	-0.0800	377	18.85	-0.0875
178	8.90	-0.0748	378	18.90	-0.0841
179	8.95	-0.0693	379	18.95	-0.0794
180	9.00	-0.0635	380	19.00	-0.0735
181	9.05	-0.0575	381	19.05	-0.0664
182	9.10	-0.0513	382	19.10	-0.0582
183	9.15	-0.0448	383	19.15	-0.0492
184	9.20	-0.0378	384	19.20	-0.0394
185	9.25	-0.0301	385	19.25	-0.0289
186	9.30	-0.0218	386	19.30	-0.0180
187	9.35	-0.0127	387	19.35	-0.0068
188	9.40	-0.0028	388	19.40	0.0045
189	9.45	0.0078	389	19.45	0.0158

(continued)

Table 2.1 (continued)

i	t	y	i	t	y
190	9.50	0.0190	390	19.50	0.0268
191	9.55	0.0305	391	19.55	0.0373
192	9.60	0.0420	392	19.60	0.0473
193	9.65	0.0532	393	19.65	0.0566
194	9.70	0.0637	394	19.70	0.0649
195	9.75	0.0733	395	19.75	0.0723
196	9.80	0.0815	396	19.80	0.0785
197	9.85	0.0880	397	19.85	0.0834
198	9.90	0.0927	398	19.90	0.0871
199	9.95	0.0955	399	19.95	0.0893
200	10.00	0.0962	400	20.00	0.0902

Table 2.2 *Typical* table for frequency content of driven damped oscillator obtained using the data of Table 2.1 in Program number 2.1. Here $I = \sqrt{(-1)}$, Abs stands for absolute value of the complex number for $A[n]$

n1	n	A[n]	Abs [A[n]]
0	0.00	-0.0091	0.0091
1	0.05	-0.00909861+0.00289636 I	0.0095
2	0.10	-0.00903001+0.00583478 I	0.0108
3	0.15	-0.00891104+0.00885936 I	0.0126
4	0.20	-0.00873419+0.0120184 I	0.0149
5	0.25	-0.00848774+0.0153672 I	0.0176
6	0.30	-0.0081542+0.0189712 I	0.0206
7	0.35	-0.00770787+0.022911 I	0.0242
8	**0.40**	0.352429+0.0229469 I	**0.3532**
9	0.45	-0.00630615+0.0322354 I	0.0328
10	0.50	-0.00520645+0.0379303 I	0.0383
11	0.55	-0.00367322+0.0446171 I	0.0448
12	0.60	-0.00147824+0.0526412 I	0.0527
13	0.65	0.00177269+0.0625049 I	0.0625
14	0.70	0.0068008+0.0749545 I	0.0753
15	0.75	0.0150205+0.091109 I	0.0923
16	0.80	0.0294431+0.112565 I	0.1164

(continued)

Table 2.2 (continued)

n1	n	A[n]	Abs [A[n]]
17	0.85	0.0570731+0.140926 I	0.1520
18	0.90	0.115014+0.173011 I	0.2078
19	0.95	0.234227+0.171529 I	0.2903
20	**1.00**	0.352942+0.0202146 I	**0.3535**
21	1.05	0.266819-0.154171 I	0.3082
22	1.10	0.147345-0.178854 I	0.2317
23	1.15	0.0832764-0.157155 I	0.1779
24	1.20	0.0506656-0.133622 I	0.1429
25	1.25	0.0326412-0.11476 I	0.1193
26	1.30	0.0217892-0.100204 I	0.1025
27	1.35	0.0147793-0.0888677 I	0.0901
28	1.40	0.00999105-0.0798677 I	0.0805
29	1.45	0.00657147-0.0725776 I	0.0729
30	1.50	0.0040399-0.0665625 I	0.0667
31	1.55	0.00210965-0.0615178 I	0.0616
32	1.60	0.000601331-0.057226 I	0.0572
33	1.65	-0.000601898-0.053529 I	0.0535
34	1.70	-0.00157882-0.0503094 I	0.0503
35	1.75	-0.00238413-0.0474785 I	0.0475
36	1.80	-0.0030568-0.0449682 I	0.0451
37	1.85	-0.00362522-0.0427254 I	0.0429
38	1.90	-0.00411046-0.0407081 I	0.0409
39	1.95	-0.00452849-0.0388827 I	0.0391
40	2.00	-0.00489154-0.0372218 I	0.0375
41	2.05	-0.00520915-0.0357032 I	0.0361
42	2.10	-0.00548886-0.0343086 I	0.0347
43	2.15	-0.00573665-0.0330225 I	0.0335
44	2.20	-0.00595737-0.0318321 I	0.0324
45	2.25	-0.00615495-0.0307264 I	0.0313
46	2.30	-0.00633264-0.0296963 I	0.0304
47	2.35	-0.00649311-0.0287337 I	0.0295
48	2.40	-0.00663859-0.0278317 I	0.0286
49	2.45	-0.00677096-0.0269845 I	0.0278

(continued)

Table 2.2 (continued)

$n1$	n	$A[n]$	$Abs\,[A[n]]$
50	2.50	-0.0068918-0.0261868 I	0.0271
51	2.55	-0.00700246-0.025434 I	0.0264
52	2.60	-0.00710409-0.0247223 I	0.0257
53	2.65	-0.00719768-0.0240481 I	0.0251
54	2.70	-0.00728408-0.0234082 I	0.0245
55	2.75	-0.00736405-0.0227999 I	0.0240
56	2.80	-0.00743822-0.0222207 I	0.0234
57	2.85	-0.00750715-0.0216683 I	0.0229
58	2.90	-0.00757136-0.0211408 I	0.0225
59	2.95	-0.00763126-0.0206364 I	0.0220
60	3.00	-0.00768726-0.0201534 I	0.0216
61	3.05	-0.00773969-0.0196904 I	0.0212
62	3.10	-0.00778886-0.019246 I	0.0208
63	3.15	-0.00783504-0.0188191 I	0.0204
64	3.20	-0.00787848-0.0184084 I	0.0200
65	3.25	-0.00791939-0.018013 I	0.0197
66	3.30	-0.00795798-0.0176319 I	0.0193
67	3.35	-0.00799441-0.0172643 I	0.0190
68	3.40	-0.00802886-0.0169094 I	0.0187
69	3.45	-0.00806146-0.0165665 I	0.0184
70	3.50	-0.00809236-0.0162349 I	0.0181
71	3.55	-0.00812166-0.0159139 I	0.0179
72	3.60	-0.00814948-0.015603 I	0.0176
73	3.65	-0.00817593-0.0153017 I	0.0173
74	3.70	-0.00820108-0.0150095 I	0.0171
75	3.75	-0.00822503-0.0147258 I	0.0169
76	3.80	-0.00824785-0.0144503 I	0.0166
77	3.85	-0.00826961-0.0141826 I	0.0164
78	3.90	-0.00829038-0.0139223 I	0.0162
79	3.95	-0.00831022-0.013669 I	0.0160
80	4.00	-0.00832919-0.0134225 I	0.0158
81	4.05	-0.00834733-0.0131823 I	0.0156
82	4.10	-0.00836469-0.0129482 I	0.0154
83	4.15	-0.00838132-0.01272 I	0.0152
84	4.20	-0.00839726-0.0124974 I	0.0151

(continued)

Table 2.2 (continued)

n1	n	A[n]	Abs [A[n]]
85	4.25	-0.00841254-0.0122801 I	0.0149
86	4.30	-0.00842721-0.0120679 I	0.0147
87	4.35	-0.00844129-0.0118606 I	0.0146
88	4.40	-0.00845481-0.011658 I	0.0144
89	4.45	-0.00846781-0.01146 I	0.0142
90	4.50	-0.00848031-0.0112662 I	0.0141
91	4.55	-0.00849234-0.0110766 I	0.0140
92	4.60	-0.00850391-0.0108909 I	0.0138
93	4.65	-0.00851506-0.0107091 I	0.0137
94	4.70	-0.0085258-0.010531 I	0.0135
95	4.75	-0.00853615-0.0103564 I	0.0134
96	4.80	-0.00854613-0.0101853 I	0.0133
97	4.85	-0.00855576-0.0100174 I	0.0132
98	4.90	-0.00856505-0.00985269 I	0.0131
99	4.95	-0.00857401-0.00969104 I	0.0129
100	5.00	-0.00858267-0.00953232 I	0.0128
101	5.05	-0.00859104-0.00937644 I	0.0127
102	5.10	-0.00859912-0.0092233 I	0.0126
103	5.15	-0.00860694-0.00907281 I	0.0125
104	5.20	-0.0086145-0.00892486 I	0.0124
105	5.25	-0.00862181-0.00877939 I	0.0123
106	5.30	-0.00862888-0.0086363 I	0.0122
107	5.35	-0.00863573-0.00849551 I	0.0121
108	5.40	-0.00864235-0.00835695 I	0.0120
109	5.45	-0.00864877-0.00822054 I	0.0119
110	5.50	-0.00865499-0.00808622 I	0.0118
111	5.55	-0.00866101-0.00795392 I	0.0118
112	5.60	-0.00866685-0.00782357 I	0.0117
113	5.65	-0.00867251-0.00769512 I	0.0116
114	5.70	-0.00867799-0.00756849 I	0.0115
115	5.75	-0.00868331-0.00744365 I	0.0114
116	5.80	-0.00868847-0.00732052 I	0.0114
117	5.85	-0.00869348-0.00719905 I	0.0113
118	5.90	-0.00869833-0.00707921 I	0.0112
119	5.95	-0.00870305-0.00696093 I	0.0111

(continued)

Table 2.2 (continued)

n1	n	A[n]	Abs [A[n]]
120	6.00	-0.00870762-0.00684417 I	0.0111
121	6.05	-0.00871206-0.00672888 I	0.0110
122	6.10	-0.00871638-0.00661503 I	0.0109
123	6.15	-0.00872056-0.00650256 I	0.0109
124	6.20	-0.00872463-0.00639144 I	0.0108
125	6.25	-0.00872858-0.00628163 I	0.0108
126	6.30	-0.00873242-0.00617308 I	0.0107
127	6.35	-0.00873615-0.00606577 I	0.0106
128	6.40	-0.00873977-0.00595965 I	0.0106
129	6.45	-0.00874329-0.00585469 I	0.0105
130	6.50	-0.00874671-0.00575086 I	0.0105
131	6.55	-0.00875003-0.00564813 I	0.0104
132	6.60	-0.00875326-0.00554646 I	0.0104
133	6.65	-0.0087564-0.00544583 I	0.0103
134	6.70	-0.00875945-0.0053462 I	0.0103
135	6.75	-0.00876242-0.00524754 I	0.0102
136	6.80	-0.0087653-0.00514983 I	0.0102
137	6.85	-0.0087681-0.00505304 I	0.0101
138	6.90	-0.00877083-0.00495715 I	0.0101
139	6.95	-0.00877348-0.00486213 I	0.0100
140	7.00	-0.00877605-0.00476795 I	0.0100
141	7.05	-0.00877855-0.00467459 I	0.0099
142	7.10	-0.00878099-0.00458202 I	0.0099
143	7.15	-0.00878335-0.00449024 I	0.0099
144	7.20	-0.00878565-0.0043992 I	0.0098
145	7.25	-0.00878788-0.0043089 I	0.0098
146	7.30	-0.00879005-0.0042193 I	0.0098
147	7.35	-0.00879215-0.0041304 I	0.0097
148	7.40	-0.0087942-0.00404216 I	0.0097
149	7.45	-0.00879619-0.00395457 I	0.0096
150	7.50	-0.00879812-0.00386762 I	0.0096
151	7.55	-0.00879999-0.00378128 I	0.0096
152	7.60	-0.00880181-0.00369553 I	0.0095
153	7.65	-0.00880358-0.00361036 I	0.0095

(continued)

Table 2.2 (continued)

$n1$	n	$A[n]$	$Abs\,[A[n]]$
154	7.70	-0.00880529-0.00352575 I	0.0095
155	7.75	-0.00880696-0.00344168 I	0.0095
156	7.80	-0.00880857-0.00335814 I	0.0094
157	7.85	-0.00881013-0.00327511 I	0.0094
158	7.90	-0.00881165-0.00319257 I	0.0094
159	7.95	-0.00881311-0.00311051 I	0.0093
160	8.00	-0.00881454-0.00302892 I	0.0093
161	8.05	-0.00881591-0.00294777 I	0.0093
162	8.10	-0.00881724-0.00286706 I	0.0093
163	8.15	-0.00881853-0.00278678 I	0.0092
164	8.20	-0.00881978-0.00270689 I	0.0092
165	8.25	-0.00882098-0.0026274 I	0.0092
166	8.30	-0.00882214-0.00254829 I	0.0092
167	8.35	-0.00882327-0.00246955 I	0.0092
168	8.40	-0.00882435-0.00239116 I	0.0091
169	8.45	-0.00882539-0.00231311 I	0.0091
170	8.50	-0.00882639-0.00223538 I	0.0091
171	8.55	-0.00882736-0.00215797 I	0.0091
172	8.60	-0.00882829-0.00208087 I	0.0091
173	8.65	-0.00882918-0.00200406 I	0.0091
174	8.70	-0.00883003-0.00192752 I	0.0090
175	8.75	-0.00883085-0.00185126 I	0.0090
176	8.80	-0.00883163-0.00177525 I	0.0090
177	8.85	-0.00883238-0.00169948 I	0.0090
178	8.90	-0.00883309-0.00162395 I	0.0090
179	8.95	-0.00883377-0.00154864 I	0.0090
180	9.00	-0.00883441-0.00147354 I	0.0090
181	9.05	-0.00883502-0.00139864 I	0.0089
182	9.10	-0.0088356-0.00132394 I	0.0089
183	9.15	-0.00883614-0.00124941 I	0.0089
184	9.20	-0.00883666-0.00117505 I	0.0089
185	9.25	-0.00883714-0.00110085 I	0.0089
186	9.30	-0.00883758-0.0010268 I	0.0089
187	9.35	-0.008838-0.000952884 I	0.0089
188	9.40	-0.00883838-0.000879097 I	0.0089

(continued)

Table 2.2 (continued)

$n1$	n	$A[n]$	$Abs\,[A[n]]$
189	9.45	-0.00883873-0.000805428 I	0.0089
190	9.50	-0.00883905-0.000731867 I	0.0089
191	9.55	-0.00883934-0.000658403 I	0.0089
192	9.60	-0.0088396-0.000585027 I	0.0089
193	9.65	-0.00883983-0.000511729 I	0.0089
194	9.70	-0.00884003-0.000438498 I	0.0089
195	9.75	-0.0088402-0.000365327 I	0.0088
196	9.80	-0.00884033-0.000292203 I	0.0088
197	9.85	-0.00884044-0.000219118 I	0.0088
198	9.90	-0.00884051-0.000146063 I	0.0088
199	9.95	-0.00884056-0.0000730266 I	0.0088
200	10.00	-0.00884057+1.56068*10^-15 I	0.0088
201	10.05	-0.00884056+0.0000730266 I	0.0088
202	10.10	-0.00884051+0.000146063 I	0.0088
203	10.15	-0.00884044+0.000219118 I	0.0088
204	10.20	-0.00884033+0.000292203 I	0.0088
205	10.25	-0.0088402+0.000365327 I	0.0088
206	10.30	-0.00884003+0.000438498 I	0.0089
207	10.35	-0.00883983+0.000511729 I	0.0089
208	10.40	-0.0088396+0.000585027 I	0.0089
209	10.45	-0.00883934+0.000658403 I	0.0089
210	10.50	-0.00883905+0.000731867 I	0.0089
211	10.55	-0.00883873+0.000805428 I	0.0089
212	10.60	-0.00883838+0.000879097 I	0.0089
213	10.65	-0.008838+0.000952884 I	0.0089
214	10.70	-0.00883758+0.0010268 I	0.0089
215	10.75	-0.00883714+0.00110085 I	0.0089
216	10.80	-0.00883666+0.00117505 I	0.0089
217	10.85	-0.00883614+0.00124941 I	0.0089
218	10.90	-0.0088356+0.00132394 I	0.0089
219	10.95	-0.00883502+0.00139864 I	0.0089
220	11.00	-0.00883441+0.00147354 I	0.0090
221	11.05	-0.00883377+0.00154864 I	0.0090
222	11.10	-0.00883309+0.00162395 I	0.0090
223	11.15	-0.00883238+0.00169948 I	0.0090

(continued)

Table 2.2 (continued)

n1	n	A[n]	Abs [A[n]]
224	11.20	-0.00883163+0.00177525 I	0.0090
225	11.25	-0.00883085+0.00185126 I	0.0090
226	11.30	-0.00883003+0.00192752 I	0.0090
227	11.35	-0.00882918+0.00200406 I	0.0091
228	11.40	-0.00882829+0.00208087 I	0.0091
229	11.45	-0.00882736+0.00215797 I	0.0091
230	11.50	-0.00882639+0.00223538 I	0.0091
231	11.55	-0.00882539+0.00231311 I	0.0091
232	11.60	-0.00882435+0.00239116 I	0.0091
233	11.65	-0.00882327+0.00246955 I	0.0092
234	11.70	-0.00882214+0.00254829 I	0.0092
235	11.75	-0.00882098+0.0026274 I	0.0092
236	11.80	-0.00881978+0.00270689 I	0.0092
237	11.85	-0.00881853+0.00278678 I	0.0092
238	11.90	-0.00881724+0.00286706 I	0.0093
239	11.95	-0.00881591+0.00294777 I	0.0093
240	12.00	-0.00881454+0.00302892 I	0.0093
241	12.05	-0.00881311+0.00311051 I	0.0093
242	12.10	-0.00881165+0.00319257 I	0.0094
243	12.15	-0.00881013+0.00327511 I	0.0094
244	12.20	-0.00880857+0.00335814 I	0.0094
245	12.25	-0.00880696+0.00344168 I	0.0095
246	12.30	-0.00880529+0.00352575 I	0.0095
247	12.35	-0.00880358+0.00361036 I	0.0095
248	12.40	-0.00880181+0.00369553 I	0.0095
249	12.45	-0.00879999+0.00378128 I	0.0096
250	12.50	-0.00879812+0.00386762 I	0.0096
251	12.55	-0.00879619+0.00395457 I	0.0096
252	12.60	-0.0087942+0.00404216 I	0.0097
253	12.65	-0.00879215+0.0041304 I	0.0097
254	12.70	-0.00879005+0.0042193 I	0.0098
255	12.75	-0.00878788+0.0043089 I	0.0098
256	12.80	-0.00878565+0.0043992 I	0.0098
257	12.85	-0.00878335+0.00449024 I	0.0099

(continued)

Table 2.2 (continued)

n1	n	A[n]	Abs [A[n]]
258	12.90	-0.00878099+0.00458202 I	0.0099
259	12.95	-0.00877855+0.00467459 I	0.0099
260	13.00	-0.00877605+0.00476795 I	0.0100
261	13.05	-0.00877348+0.00486213 I	0.0100
262	13.10	-0.00877083+0.00495715 I	0.0101
263	13.15	-0.0087681+0.00505304 I	0.0101
264	13.20	-0.0087653+0.00514983 I	0.0102
265	13.25	-0.00876242+0.00524754 I	0.0102
266	13.30	-0.00875945+0.0053462 I	0.0103
267	13.35	-0.0087564+0.00544583 I	0.0103
268	13.40	-0.00875326+0.00554646 I	0.0104
269	13.45	-0.00875003+0.00564813 I	0.0104
270	13.50	-0.00874671+0.00575086 I	0.0105
271	13.55	-0.00874329+0.00585469 I	0.0105
272	13.60	-0.00873977+0.00595965 I	0.0106
273	13.65	-0.00873615+0.00606577 I	0.0106
274	13.70	-0.00873242+0.00617308 I	0.0107
275	13.75	-0.00872858+0.00628163 I	0.0108
276	13.80	-0.00872463+0.00639144 I	0.0108
277	13.85	-0.00872056+0.00650256 I	0.0109
278	13.90	-0.00871638+0.00661503 I	0.0109
279	13.95	-0.00871206+0.00672888 I	0.0110
280	14.00	-0.00870762+0.00684417 I	0.0111
281	14.05	-0.00870305+0.00696093 I	0.0111
282	14.10	-0.00869833+0.00707921 I	0.0112
283	14.15	-0.00869348+0.00719905 I	0.0113
284	14.20	-0.00868847+0.00732052 I	0.0114
285	14.25	-0.00868331+0.00744365 I	0.0114
286	14.30	-0.00867799+0.00756849 I	0.0115
287	14.35	-0.00867251+0.00769512 I	0.0116
288	14.40	-0.00866685+0.00782357 I	0.0117

(continued)

Table 2.2 (continued)

n1	n	A[n]	Abs [A[n]]
289	14.45	-0.00866101+0.00795392 I	0.0118
290	14.50	-0.00865499+0.00808622 I	0.0118
291	14.55	-0.00864877+0.00822054 I	0.0119
292	14.60	-0.00864235+0.00835695 I	0.0120
293	14.65	-0.00863573+0.00849551 I	0.0121
294	14.70	-0.00862888+0.0086363 I	0.0122
295	14.75	-0.00862181+0.00877939 I	0.0123
296	14.80	-0.0086145+0.00892486 I	0.0124
297	14.85	-0.00860694+0.00907281 I	0.0125
298	14.90	-0.00859912+0.0092233 I	0.0126
299	14.95	-0.00859104+0.00937644 I	0.0127
300	15.00	-0.00858267+0.00953232 I	0.0128
301	15.05	-0.00857401+0.00969104 I	0.0129
302	15.10	-0.00856505+0.00985269 I	0.0131
303	15.15	-0.00855576+0.0100174 I	0.0132
304	15.20	-0.00854613+0.0101853 I	0.0133
305	15.25	-0.00853615+0.0103564 I	0.0134
306	15.30	-0.0085258+0.010531 I	0.0135
307	15.35	-0.00851506+0.0107091 I	0.0137
308	15.40	-0.00850391+0.0108909 I	0.0138
309	15.45	-0.00849234+0.0110766 I	0.0140
310	15.50	-0.00848031+0.0112662 I	0.0141
311	15.55	-0.00846781+0.01146 I	0.0142
312	15.60	-0.00845481+0.011658 I	0.0144
313	15.65	-0.00844129+0.0118606 I	0.0146
314	15.70	-0.00842721+0.0120679 I	0.0147
315	15.75	-0.00841254+0.0122801 I	0.0149
316	15.80	-0.00839726+0.0124974 I	0.0151
317	15.85	-0.00838132+0.01272 I	0.0152
318	15.90	-0.00836469+0.0129482 I	0.0154

(continued)

Table 2.2 (continued)

n1	n	A[n]	Abs [A[n]]
319	15.95	-0.00834733+0.0131823 I	0.0156
320	16.00	-0.00832919+0.0134225 I	0.0158
321	16.05	-0.00831022+0.013669 I	0.0160
322	16.10	-0.00829038+0.0139223 I	0.0162
323	16.15	-0.00826961+0.0141826 I	0.0164
324	16.20	-0.00824785+0.0144503 I	0.0166
325	16.25	-0.00822503+0.0147258 I	0.0169
326	16.30	-0.00820108+0.0150095 I	0.0171
327	16.35	-0.00817593+0.0153017 I	0.0173
328	16.40	-0.00814948+0.015603 I	0.0176
329	16.45	-0.00812166+0.0159139 I	0.0179
330	16.50	-0.00809236+0.0162349 I	0.0181
331	16.55	-0.00806146+0.0165665 I	0.0184
332	16.60	-0.00802886+0.0169094 I	0.0187
333	16.65	-0.00799441+0.0172643 I	0.0190
334	16.70	-0.00795798+0.0176319 I	0.0193
335	16.75	-0.00791939+0.018013 I	0.0197
336	16.80	-0.00787848+0.0184084 I	0.0200
337	16.85	-0.00783504+0.0188191 I	0.0204
338	16.90	-0.00778886+0.019246 I	0.0208
339	16.95	-0.00773969+0.0196904 I	0.0212
340	17.00	-0.00768726+0.0201534 I	0.0216
341	17.05	-0.00763126+0.0206364 I	0.0220
342	17.10	-0.00757136+0.0211408 I	0.0225
343	17.15	-0.00750715+0.0216683 I	0.0229
344	17.20	-0.00743822+0.0222207 I	0.0234
345	17.25	-0.00736405+0.0227999 I	0.0240
346	17.30	-0.00728408+0.0234082 I	0.0245
347	17.35	-0.00719768+0.0240481 I	0.0251
348	17.40	-0.00710409+0.0247223 I	0.0257
349	17.45	-0.00700246+0.025434 I	0.0264
350	17.50	-0.0068918+0.0261868 I	0.0271
351	17.55	-0.00677096+0.0269845 I	0.0278
352	17.60	-0.00663859+0.0278317 I	0.0286

(continued)

Table 2.2   (continued)

n1	n	A[n]	Abs [A[n]]
353	17.65	-0.00649311+0.0287337 I	0.0295
354	17.70	-0.00633264+0.0296963 I	0.0304
355	17.75	-0.00615495+0.0307264 I	0.0313
356	17.80	-0.00595737+0.0318321 I	0.0324
357	17.85	-0.00573665+0.0330225 I	0.0335
358	17.90	-0.00548886+0.0343086 I	0.0347
359	17.95	-0.00520915+0.0357032 I	0.0361
360	18.00	-0.00489154+0.0372218 I	0.0375
361	18.05	-0.00452849+0.0388827 I	0.0391
362	18.10	-0.00411046+0.0407081 I	0.0409
363	18.15	-0.00362522+0.0427254 I	0.0429
364	18.20	-0.0030568+0.0449682 I	0.0451
365	18.25	-0.00238413+0.0474785 I	0.0475
366	18.30	-0.00157882+0.0503094 I	0.0503
367	18.35	-0.000601898+0.053529 I	0.0535
368	18.40	0.000601331+0.057226 I	0.0572
369	18.45	0.00210965+0.0615178 I	0.0616
370	18.50	0.0040399+0.0665625 I	0.0667
371	18.55	0.00657147+0.0725776 I	0.0729
372	18.60	0.00999105+0.0798677 I	0.0805
373	18.65	0.0147793+0.0888677 I	0.0901
374	18.70	0.0217892+0.100204 I	0.1025
375	18.75	0.0326412+0.11476 I	0.1193
376	18.80	0.0506656+0.133622 I	0.1429
377	18.85	0.0832764+0.157155 I	0.1779
378	18.90	0.147345+0.178854 I	0.2317
379	18.95	0.266819+0.154171 I	0.3082
380	**19.00**	0.352942-0.0202146 I	**0.3535**
381	19.05	0.234227-0.171529 I	0.2903
382	19.10	0.115014-0.173011 I	0.2078
383	19.15	0.0570731-0.140926 I	0.1520
384	19.20	0.0294431-0.112565 I	0.1164
385	19.25	0.0150205-0.091109 I	0.0923
386	19.30	0.0068008-0.0749545 I	0.0753

(continued)

Table 2.2 (continued)

$n1$	n	$A[n]$	$Abs[A[n]]$
387	19.35	0.00177269-0.0625049 I	0.0625
388	19.40	-0.00147824-0.0526412 I	0.0527
389	19.45	-0.00367322-0.0446171 I	0.0448
390	19.50	-0.00520645-0.0379303 I	0.0383
391	19.55	-0.00630615-0.0322354 I	0.0328
392	**19.60**	0.352429-0.0229469 I	**0.3532**
393	19.65	-0.00770787-0.022911 I	0.0242
394	19.70	-0.0081542-0.0189712 I	0.0206
395	19.75	-0.00848774-0.0153672 I	0.0176
396	19.80	-0.00873419-0.0120184 I	0.0149
397	19.85	-0.00891104-0.00885936 I	0.0126
398	19.90	-0.00903001-0.00583478 I	0.0108
399	19.95	-0.00909861-0.00289636 I	0.0095
400	20.00	-0.00912103-1.54888*10^-14 I	0.0091

Key results are in bold face.

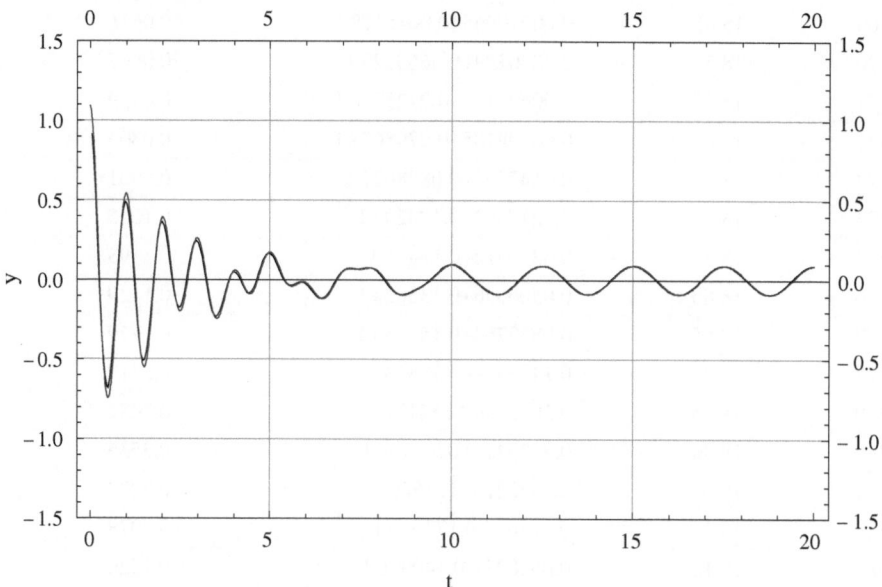

Fig. 2.1 Typical plot of displacement of driven damped oscillator as a function of time t. The parameters are $\omega_0 = 2\pi$; $\omega = 0.4\omega_0$; $\nu = \omega/(2\pi)$, a = 3; b = 1; $\omega_d = \sqrt{(\omega_0^2 - (b/2)^2)}$; $\nu_d = \omega_d/(2\pi)$, $\nu_d = 0.997$, $\nu = 0.4$. Showing analytical and numerical results together

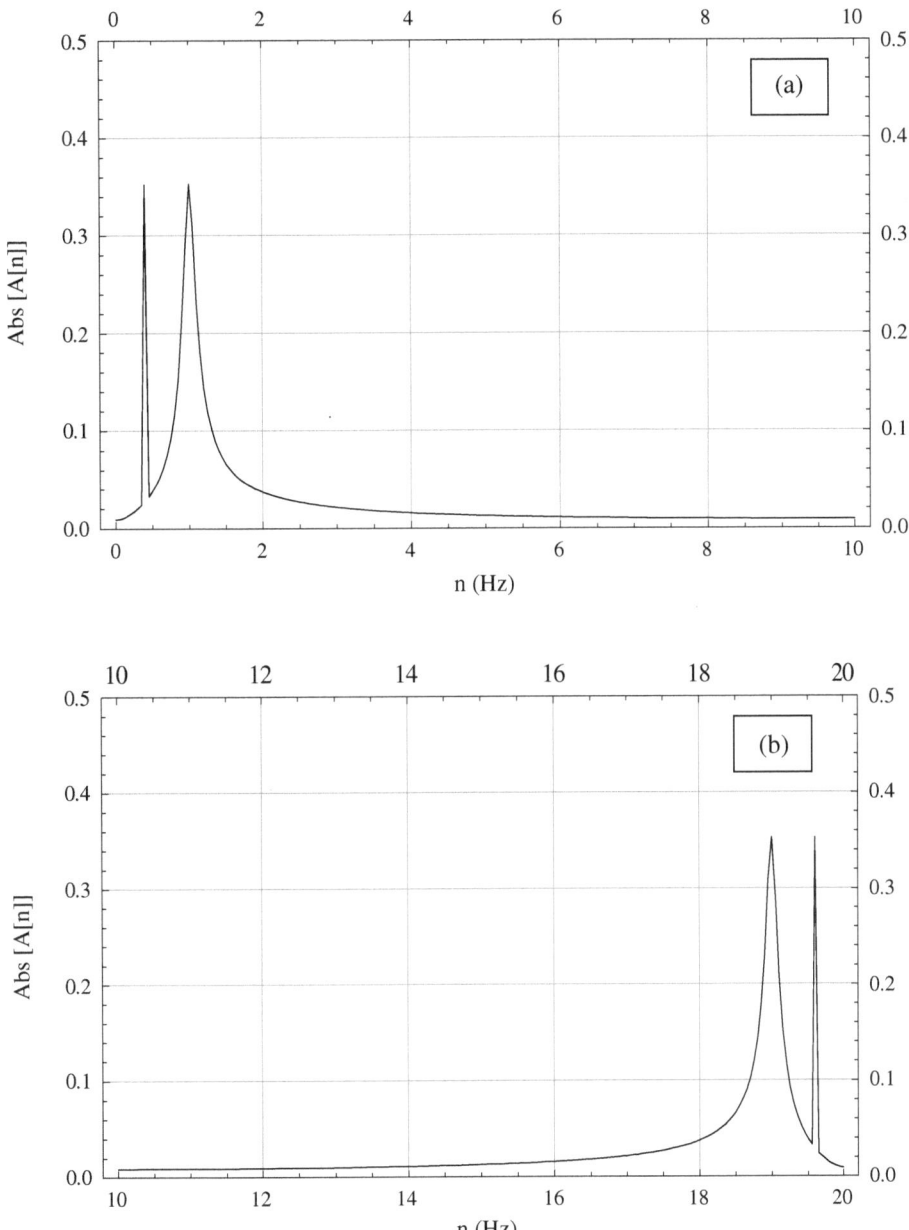

Fig. 2.2   a Showing typical frequency content in the data of driven damped oscillator. This is called power spectrum of the driven damped oscillator. The peaks at $n = 19$ and 19.6 in Fig. 2.2b are unphysical as discussed in Sect. 1.1 and hence can be ignored

Program number 2.1

```
w0=2*Pi*1.0;
w=0.4*w0;
nu=w/(2*Pi)
a=3;
b=1;
wd=Sqrt[w0^2-(b/2)^2];
nud=N[wd/(2*Pi)]


h=N[20/400];
t=0;
v=0;
y=1;
i=0;
Table[{i=i+1,t=t+h,
k1v=h*((-w0^2)*y-b*v+a*Cos[w*(t-h)]);
k2v=h*((-w0^2)*y-b*(v+k1v/2)+a*Cos[w*(t-h+h/2)]);
k3v=h*((-w0^2)*y-b*(v+k2v/2)+a*Cos[w*(t-h+h/2)]);
k4v=h*((-w0^2)*y-b*(v+k3v)+a*Cos[w*(t-h+h)]);
v=v+(k1v+2*k2v+2*k3v+k4v)/6;
k1y=h*(v);k2y=h*(v);k3y=h*(v);k4y=h*(v);
yd[i]=y=y+(k1y+2*k2y+2*k3y+k4y)/6},{i,0,399,1}];
TableForm[%,TableSpacing->{2,2},TableHeadings->{None,{"i","t","y"}}]


t=0;
v=0;
y=1;
i=0;
p1=ListLinePlot[Table[{i=i+1;t=t+h,yd[i]},{i,0,399,1}],
Frame->True,FrameLabel->{"t","y"},PlotStyle->{Black},
PlotRange->{-1.5,1.5},GridLines->Automatic,FrameTicks->All];
p2=Plot[(Exp[-b*t/2])*Cos[wd*t]+
(a/Sqrt[(w0^2-w^2)^2+(b*w)^2])*Sin[w*t-ArcTan[(w^2-w0^2)/(b*w)]],
{t,0,20},PlotRange->{-1.5,1.5},PlotStyle->{Blue}];
```

Show[p1,p2]

Z=N[Exp[-2*Pi*I/400]];

Table[{n1=n1+1,n=N[n1/20],A[n1]=h*(1/Sqrt[2*Pi])*(Sum[yd[k]*(Z^(n1*k)),
{k,1,400,1}]),Ab[n1]=Abs[A[n1]]},{n1,0-1,400-1,1}];
TableForm[%,TableSpacing->{3,3},
TableHeadings->{None,{"n1","n","A[n]","Abs [A[n]]"}}]

ListLinePlot[Table[{n1=n1+1;n=n1/20,Ab[n1]},{n1,0-1,200-1,1}],
Frame->True,FrameLabel->{"n (Hz)","Abs [A[n]]"},PlotStyle->{Black},
GridLines->Automatic,FrameTicks->All,PlotRange->{0,0.5}]

ListLinePlot[Table[{n1=n1+1;n=n1/20,Ab[n1]},{n1,200-1,400-1,1}],
Frame->True,FrameLabel->{"n (Hz)","Abs [A[n]]"},PlotStyle->{Black},
GridLines->Automatic,FrameTicks->All,PlotRange->{0,0.5}]

2.4   Motion and Power Spectrum of Driven Damped Oscillator for Various Values of Parameters

Plots for displacement y of driven damped oscillator as function of time t and corresponding power spectrum showing frequency content of the system are shown below. These plots have been obtained using Program number 2.1 for various values of parameters (Figs. 2.3, 2.4, 2.5, 2.6, 2.7, 2.8, 2.9, 2.10, 2.11, Table 2.3, Figs. 2.12, 2.13, 2.14, 2.15, 2.16, 2.17, 2.18, 2.19, 2.20, 2.21, Table 2.4 and Fig. 2.22).

We now reduce value of the damping constant b from 1 to 0.5. Hence steady state amplitude of the oscillator is reached for t > 20 s. As such, we have to use larger time interval of 30 s (rather than 20 s) to get to the steady state amplitude. We have to modify Program number 2.1. The modified program is Program number 2.2 (Figs. 2.23, 2.24, 2.25, 2.26, 2.27, 2.28, 2.29, 2.30, 2.31, Table 2.5 and Fig. 2.32).

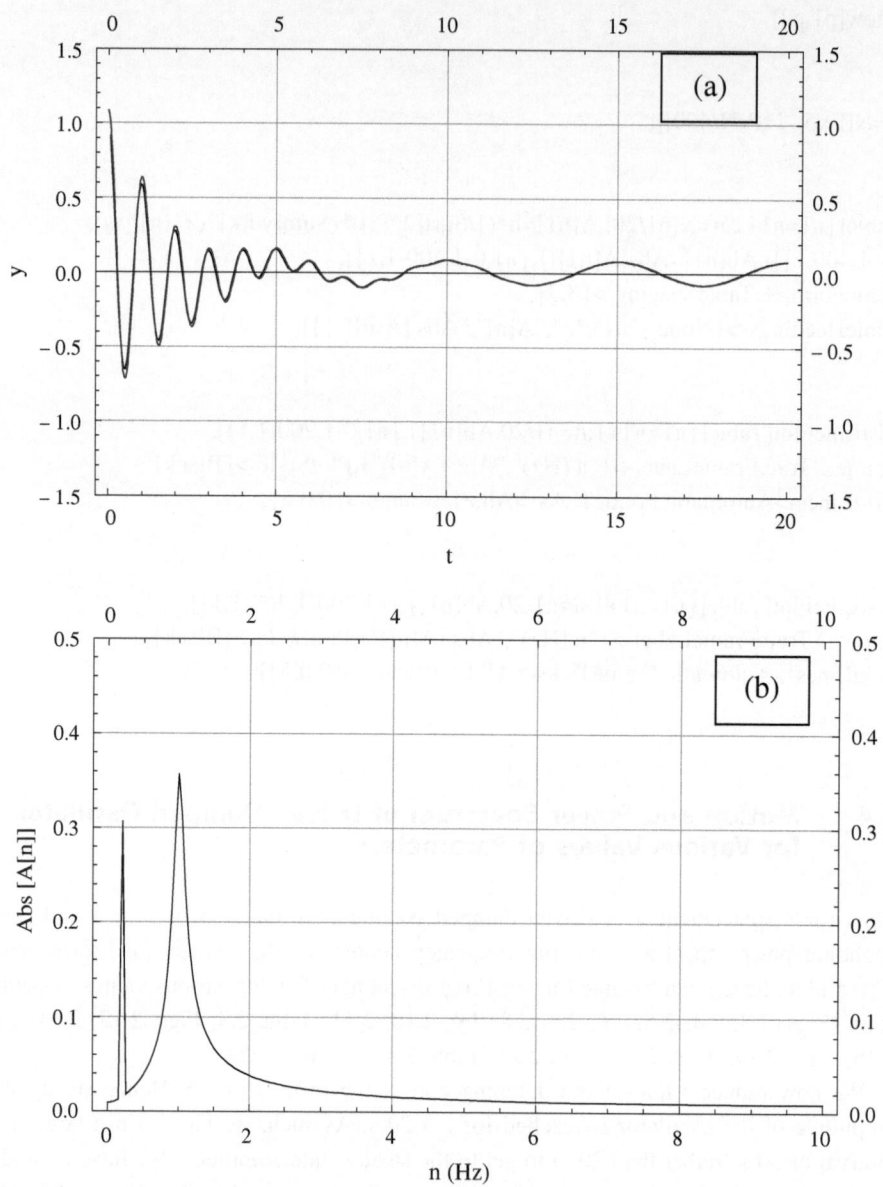

Fig. 2.3 a Shows displacement y of driven damped oscillator as a function of time t. Both analytical and numerical results are shown. **b** Shows frequency content in the data of driven damped oscillator, called power spectrum. The parameters are $\omega_0 = 2\pi$; $\omega = 0.2\omega_0$; $\nu = \omega/(2\pi)$, $a = 3$; $b = 1$; $\omega_d = \sqrt{(\omega_0^2 - (b/2)^2)}$; $\nu_d = \omega_d/(2\pi)$, $\nu_d = 0.997$, $\nu = 0.2$

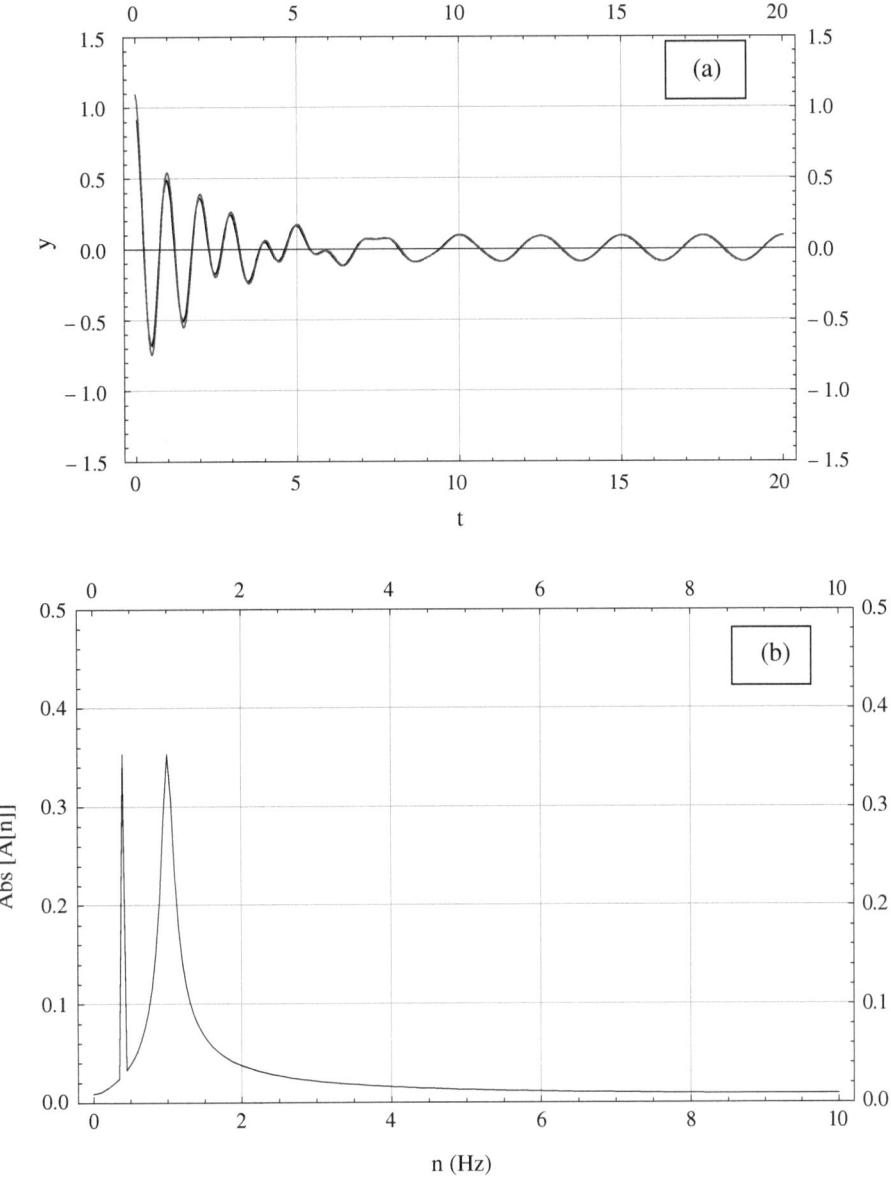

Fig. 2.4 a Shows displacement y of driven damped oscillator as a function of time t. Both analytical and numerical results are shown. **b** Shows frequency content in the data of driven damped oscillator, called power spectrum. The parameters are $\omega_0 = 2\pi$; $\omega = 0.4\omega_0$; $\nu = \omega/(2\pi)$, $a = 3$; $b = 1$; $\omega_d = \sqrt{(\omega_0^2 - (b/2)^2)}$; $\nu_d = \omega_d/(2\pi)$, $\nu_d = 0.997$, $\nu = 0.4$

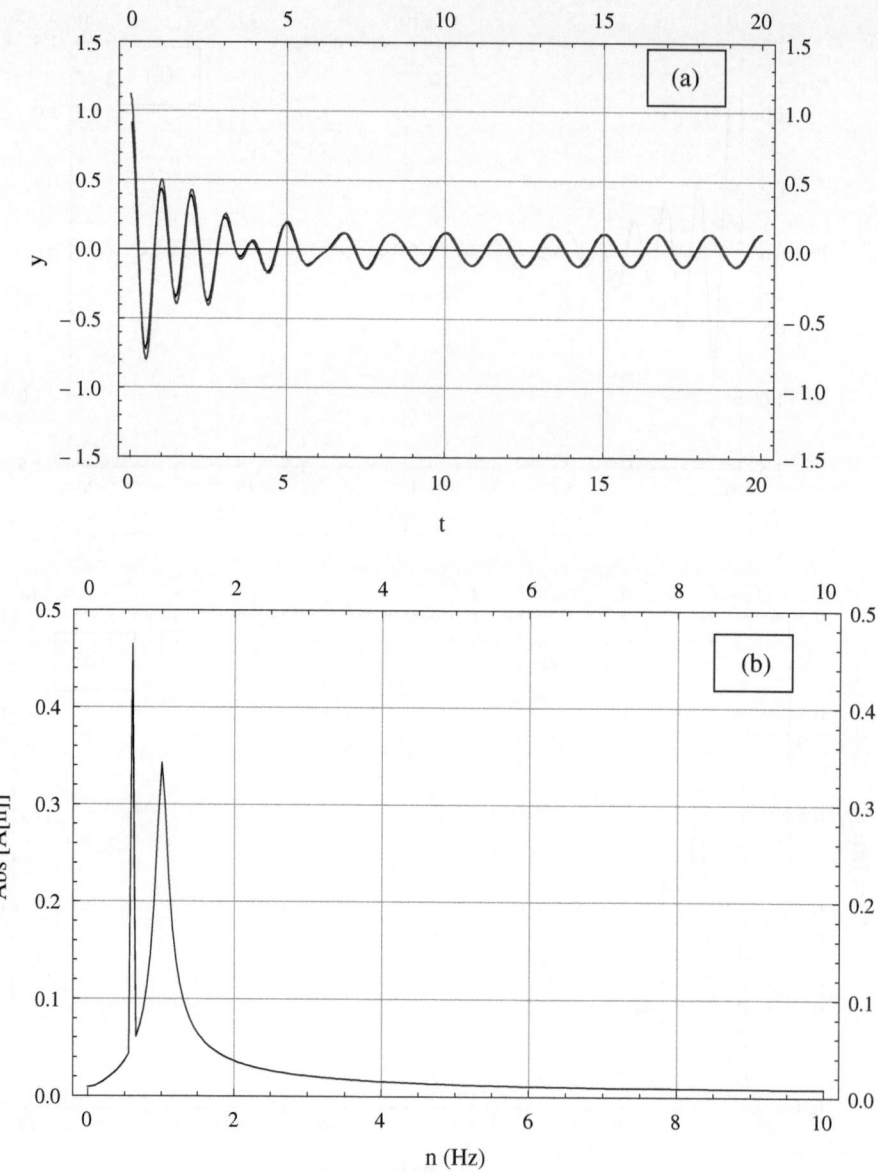

Fig. 2.5 **a** Shows displacement y of driven damped oscillator as a function of time t. Both analytical and numerical results are shown. **b** Shows frequency content in the data of driven damped oscillator, called power spectrum. The parameters are $\omega_0 = 2\pi$; $\omega = 0.6\omega_0$; $\nu = \omega/(2\pi)$, a = 3; b = 1; $\omega_d = \sqrt{(\omega_0^2 - (b/2)^2)}$; $\nu_d = \omega_d/(2\pi)$, $\nu_d = 0.997$, $\nu = 0.6$

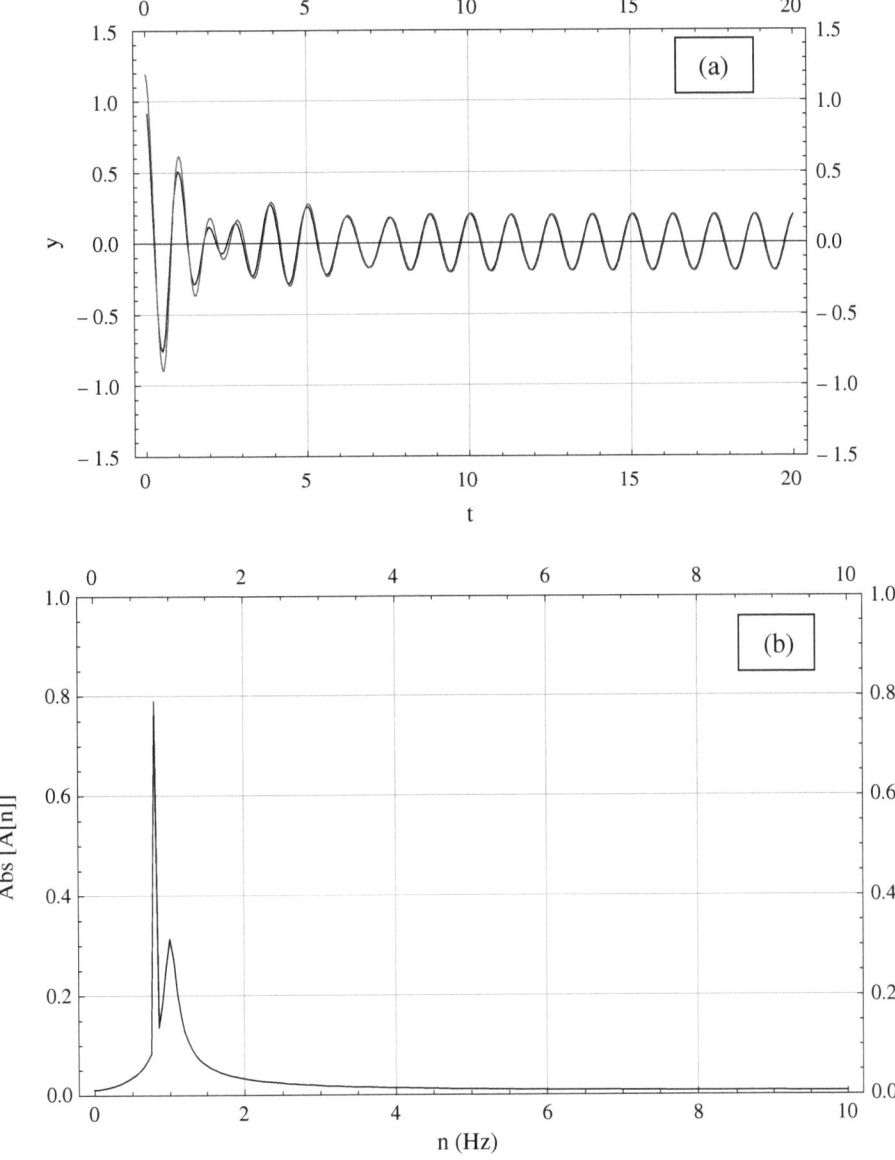

Fig. 2.6 **a** Shows displacement y of driven damped oscillator as a function of time t. Both analytical and numerical results are shown. **b** Shows frequency content in the data of driven damped oscillator, called power spectrum. The parameters are $\omega_0 = 2\pi$; $\omega = 0.8\omega_0$; $\nu = \omega/(2\pi)$, $a = 3$; $b = 1$; $\omega_d = \sqrt{(\omega_0^2 - (b/2)^2)}$; $\nu_d = \omega_d/(2\pi)$, $\nu_d = 0.997$, $\nu = 0.8$

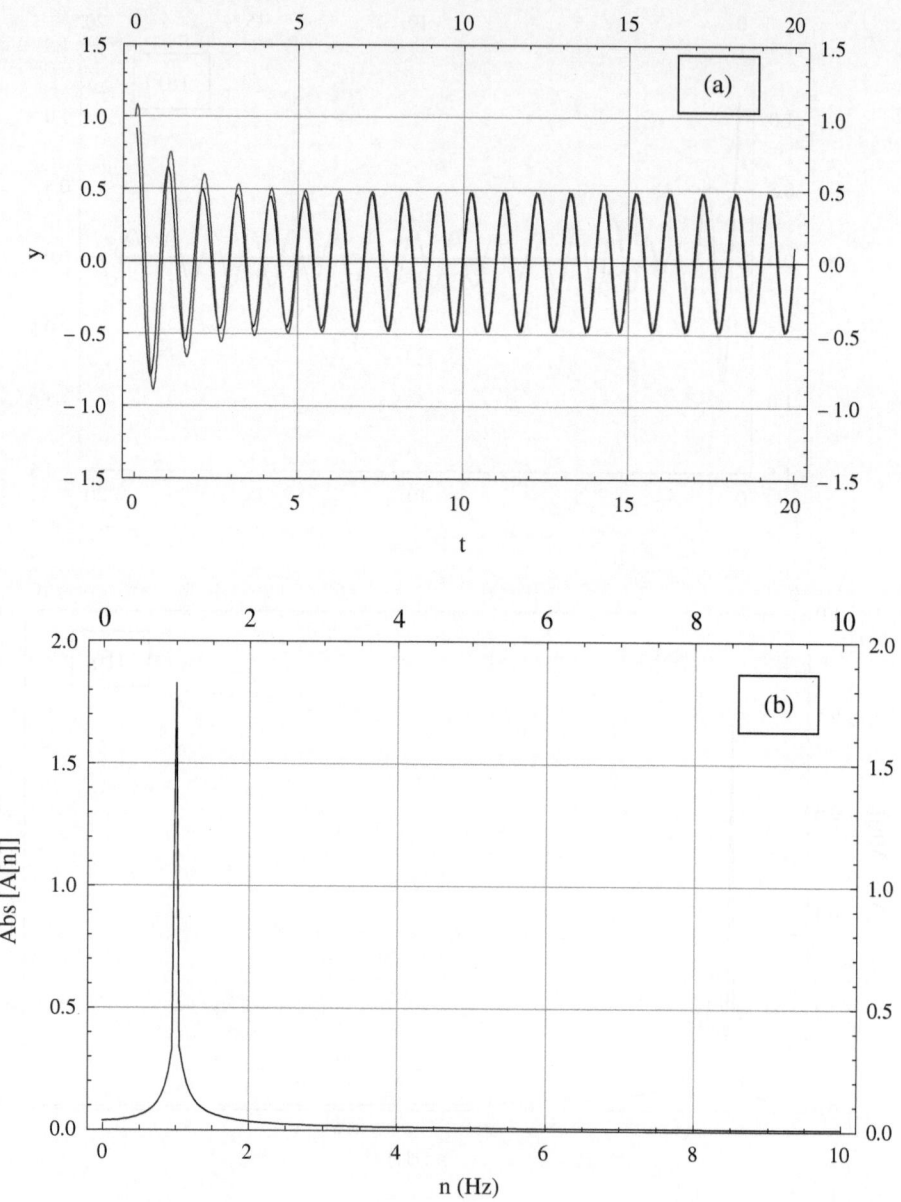

Fig. 2.7 **a** Shows displacement y of driven damped oscillator as a function of time t. Both analytical and numerical results are shown. **b** Shows frequency content in the data of driven damped oscillator, called power spectrum. The parameters are $\omega_0 = 2\pi$; $\omega = 1.0\omega_0$; $\nu = \omega/(2\pi)$, a = 3; b = 1; $\omega_d = \sqrt{(\omega_0^2 - (b/2)^2)}$; $\nu_d = \omega_d/(2\pi)$, $\nu_d = 0.997$, $\nu = 1.0$

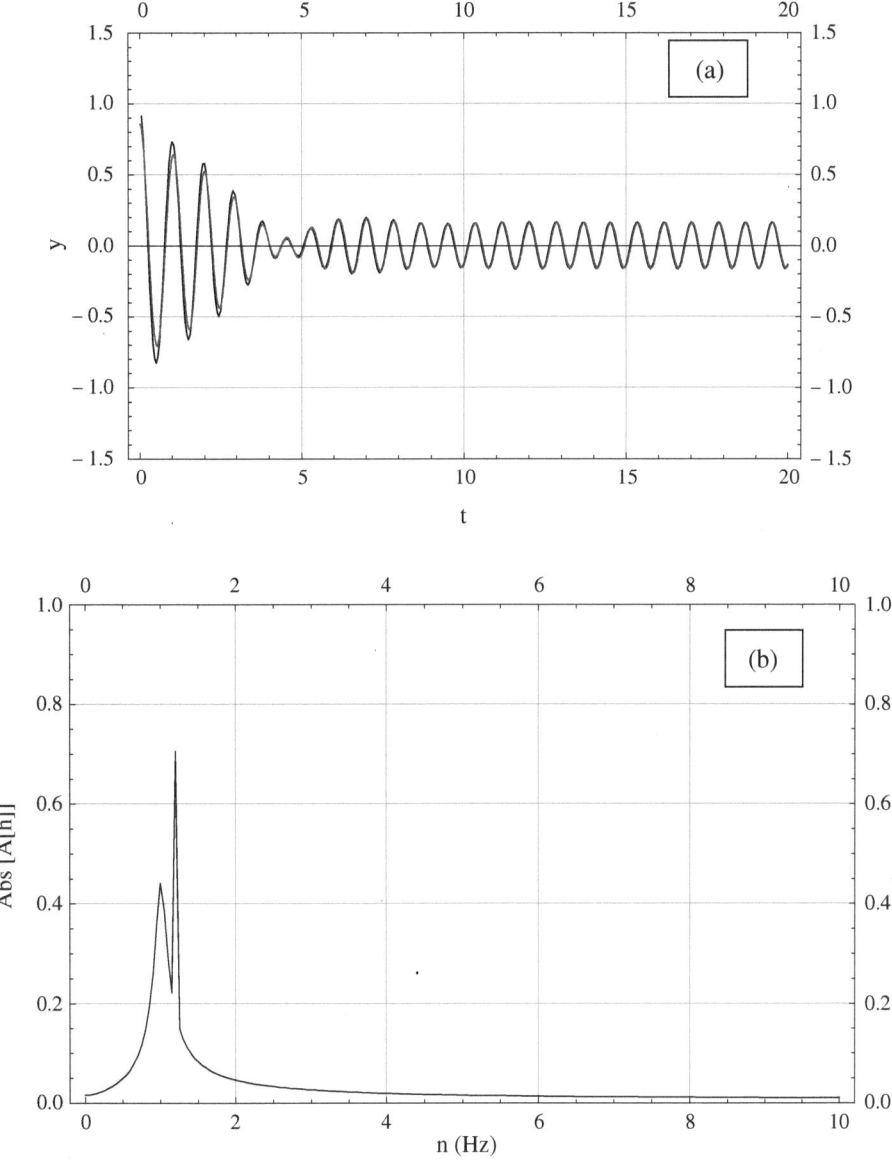

Fig. 2.8   a Shows displacement y of driven damped oscillator as a function of time t. Both analytical and numerical results are shown. **b** Shows frequency content in the data of driven damped oscillator, called power spectrum. The parameters are $\omega_0 = 2\pi$; $\omega = 1.2\omega_0$; $\nu = \omega/(2\pi)$, $a = 3$; $b = 1$; $\omega_d = \sqrt{(\omega_0^2 - (b/2)^2)}$; $\nu_d = \omega_d/(2\pi)$, $\nu_d = 0.997$, $\nu = 1.2$

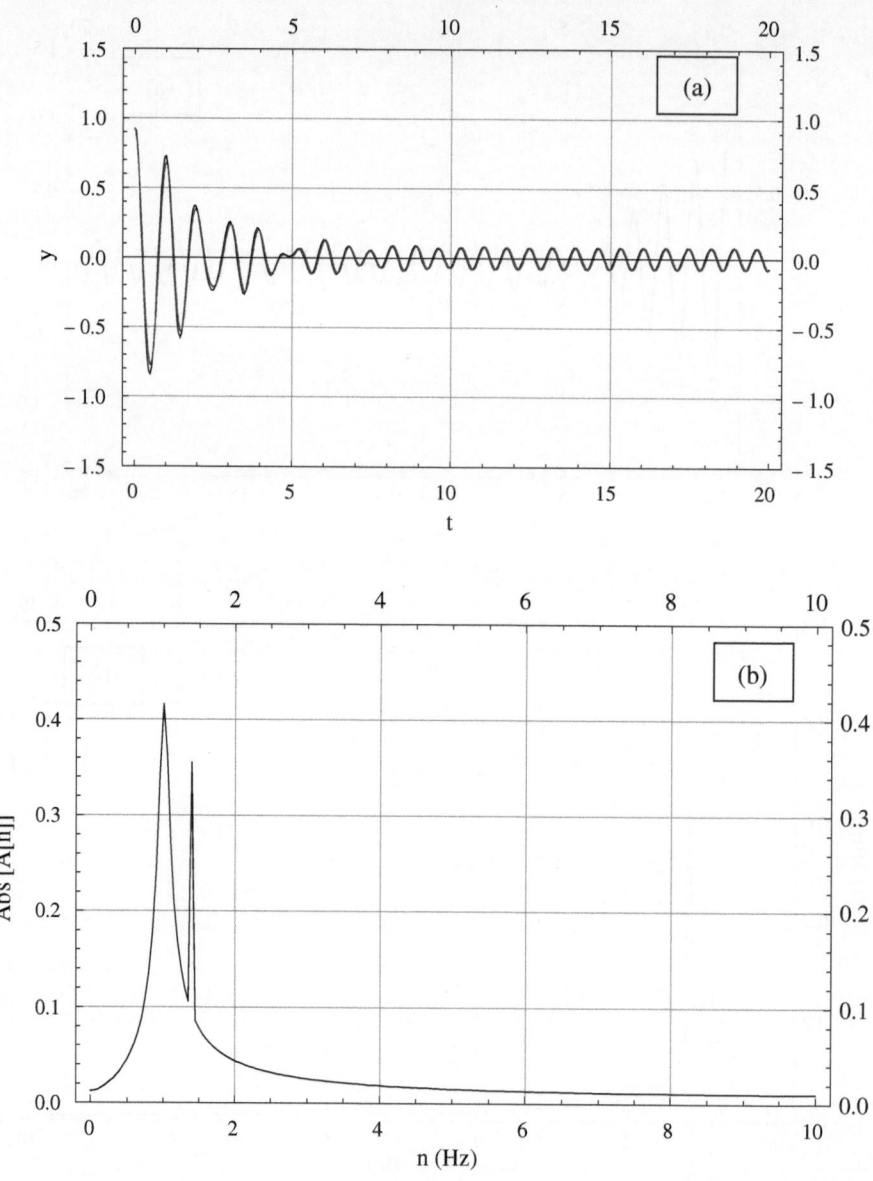

Fig. 2.9 a Shows displacement y of driven damped oscillator as a function of time t. Both analytical and numerical results are shown. **b** Shows frequency content in the data of driven damped oscillator, called power spectrum. The parameters are $\omega_0 = 2\pi$; $\omega = 1.4\omega_0$; $\nu = \omega/(2\pi)$, $a = 3$; $b = 1$; $\omega_d = \sqrt{(\omega_0^2 - (b/2)^2)}$; $\nu_d = \omega_d/(2\pi)$, $\nu_d = 0.997$, $\nu = 1.4$

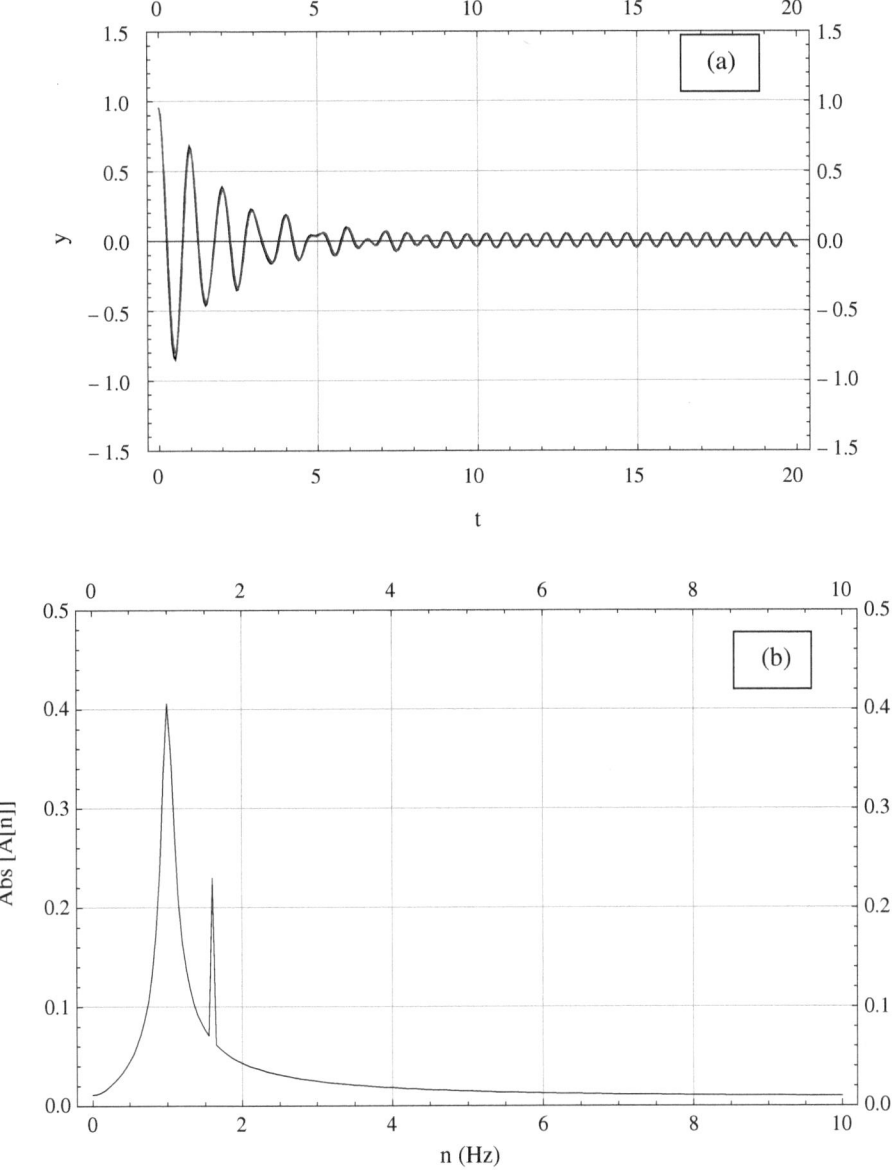

Fig. 2.10 a Shows displacement y of driven damped oscillator as a function of time t. Both analytical and numerical results are shown. **b** Shows frequency content in the data of driven damped oscillator, called power spectrum. The parameters are $\omega_0 = 2\pi$; $\omega = 1.6\omega_0$; $\nu = \omega/(2\pi)$, a = 3; b = 1; $\omega_d = \sqrt{(\omega_0^2 - (b/2)^2)}$; $\nu_d = \omega_d/(2\pi)$, $\nu_d = 0.997$, $\nu = 1.6$

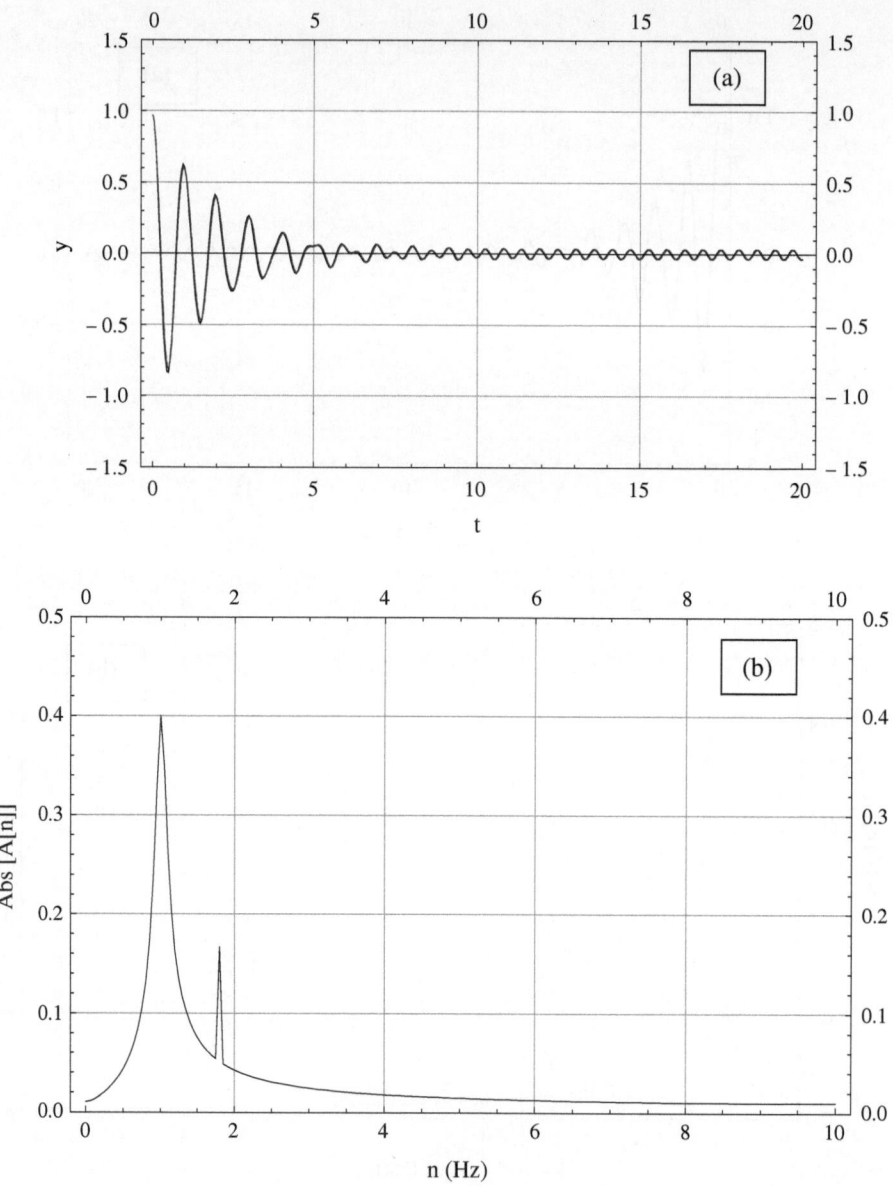

Fig. 2.11 a Shows displacement y of driven damped oscillator as a function of time t. Both analytical and numerical results are shown. **b** Shows frequency content in the data of driven damped
oscillator, called power spectrum. The parameters are $\omega_0 = 2\pi$; $\omega = 1.8\omega_0$; $\nu = \omega/(2\pi)$, $a = 3$; $b
= 1$; $\omega_d = \sqrt{(\omega_0^2 - (b/2)^2)}$; $\nu_d = \omega_d/(2\pi)$, $\nu_d = 0.997$, $\nu = 1.8$

Table 2.3 Amplitude of steady state driven oscillations as a function of $r = \omega/\omega_0$ showing resonance for $r = 1$. For the parameters $\omega_0 = 2\pi$; a = 3; b = 1 used in obtaining Figs. 2.3, 2.4, 2.5, 2.6, 2.7, 2.8, 2.9, 2.10 and 2.11

$r = \omega/\omega_0$	Amplitude
0.2	0.084
0.4	0.089
0.6	0.120
0.8	0.200
1.0	0.470
1.2	0.160
1.4	0.080
1.6	0.050
1.8	0.035

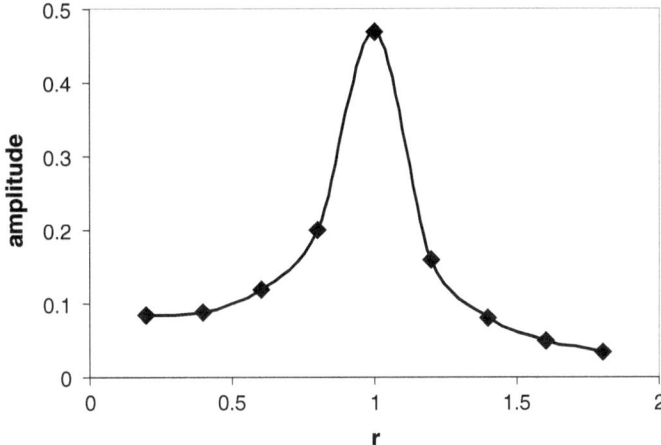

Fig. 2.12 Showing *resonance*. Amplitudes of steady state driven oscillations as function of $r = \omega/\omega_0$. For the parameters $\omega_0 = 2\pi$; a = 3; b = 1 used in obtaining Figs. 2.3, 2.4, 2.5, 2.6, 2.7, 2.8, 2.9, 2.10 and 2.11. Using Control+D command in Mathematica, we found a cursor using which we obtained the amplitudes from Figs. 2.3, 2.4, 2.5, 2.6, 2.7, 2.8, 2.9, 2.10 and 2.11 graphically

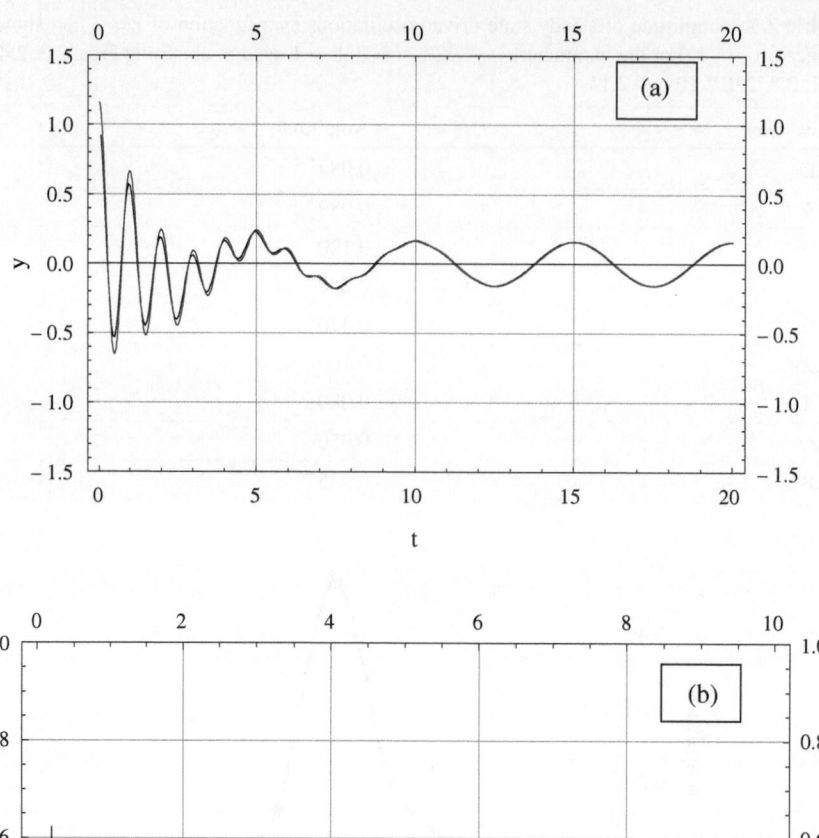

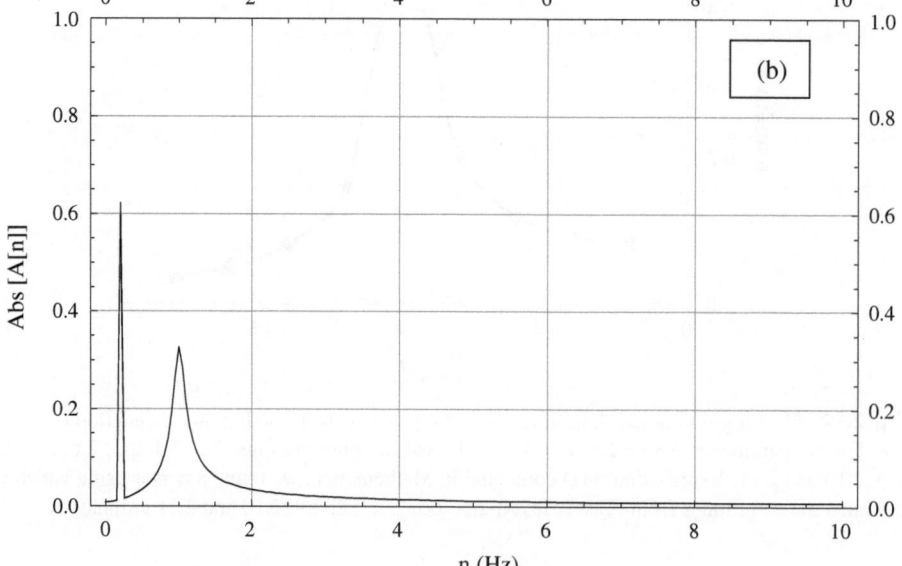

Fig. 2.13 **a** Shows displacement y of driven damped oscillator as a function of time t. Both analytical and numerical results are shown. **b** Shows frequency content in the data of driven damped oscillator, called power spectrum. The parameters are $\omega_0 = 2\pi$; $\omega = 0.2\omega_0$; $\nu = \omega/(2\pi)$, **a = 6**; b $= 1$; $\omega_d = \sqrt{(\omega_0^2 - (b/2)^2)}$; $\nu_d = \omega_d/(2\pi)$, $\nu_d = 0.997$, $\nu = 0.2$

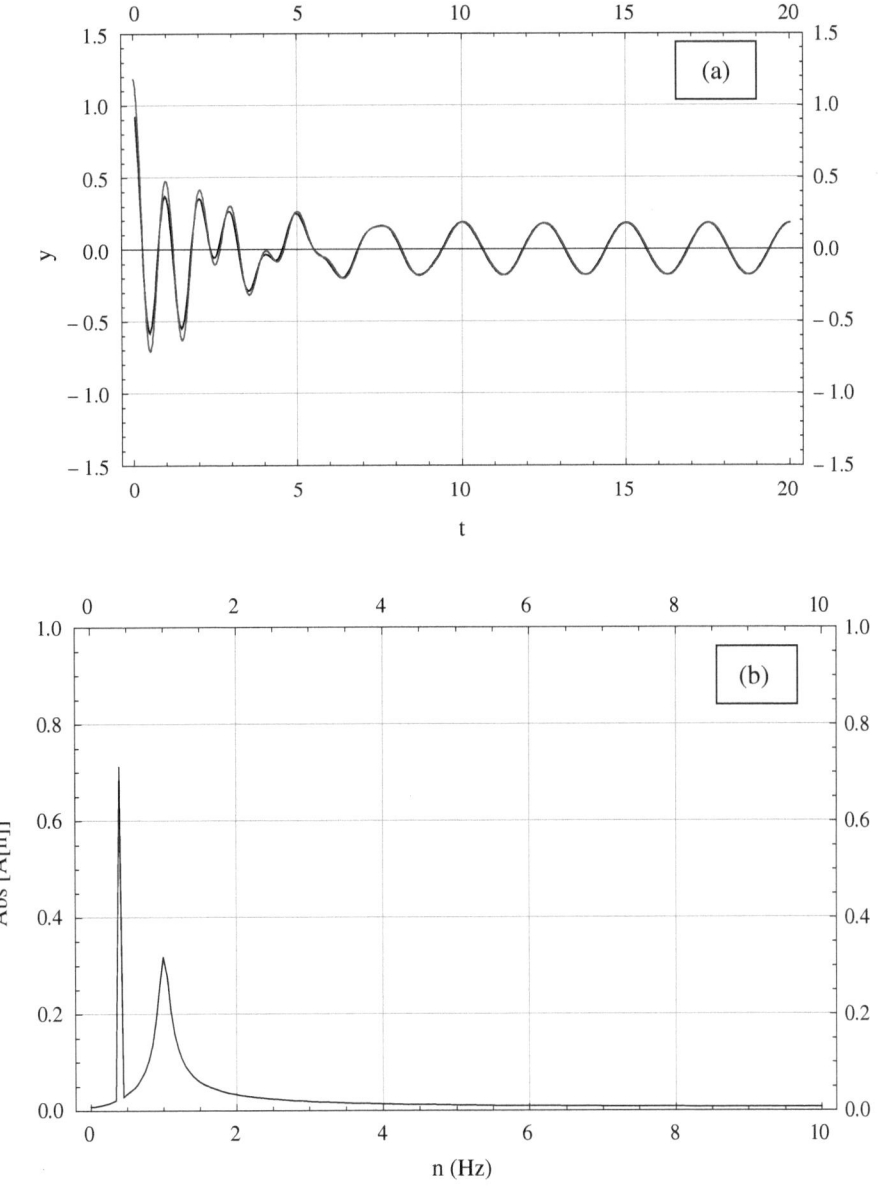

Fig. 2.14 **a** Shows displacement y of driven damped oscillator as a function of time t. Both analytical and numerical results are shown. **b** Shows frequency content in the data of driven damped
oscillator, called power spectrum. The parameters are $\omega_0 = 2\pi$; $\omega = 0.4\omega_0$; $\nu = \omega/(2\pi)$, **a** = **6**; b
= 1; $\omega_d = \sqrt{(\omega_0^2 - (b/2)^2)}$; $\nu_d = \omega_d/(2\pi)$, $\nu_d = 0.997$, $\nu = 0.4$

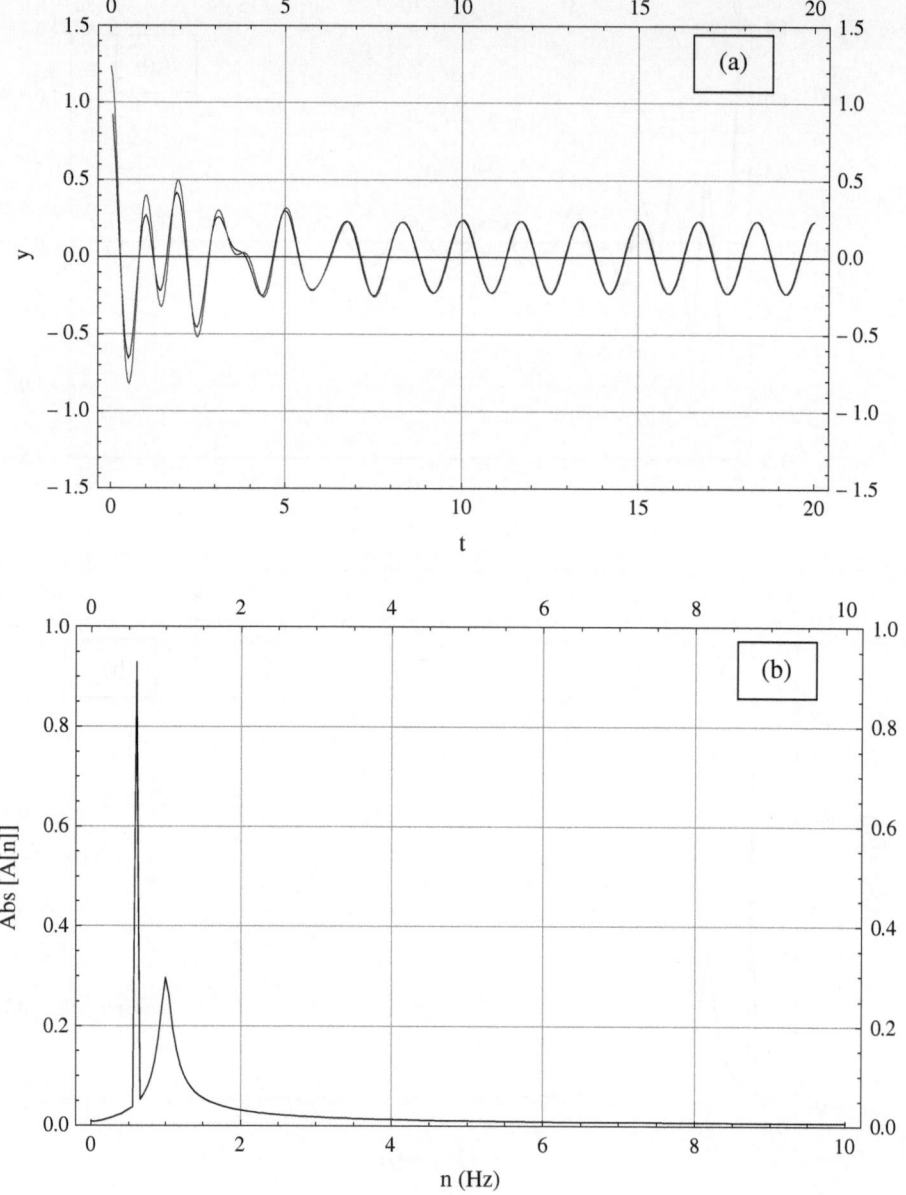

Fig. 2.15 **a** Shows displacement y of driven damped oscillator as a function of time t. Both analytical and numerical results are shown. **b** Shows frequency content in the data of driven damped oscillator, called power spectrum. The parameters are $\omega_0 = 2\pi$; $\omega = 0.6\omega_0$; $\nu = \omega/(2\pi)$, **a = 6**; b $= 1$; $\omega_d = \sqrt{(\omega_0^2 - (b/2)^2)}$; $\nu_d = \omega_d/(2\pi)$, $\nu_d = 0.997$, $\nu = 0.6$

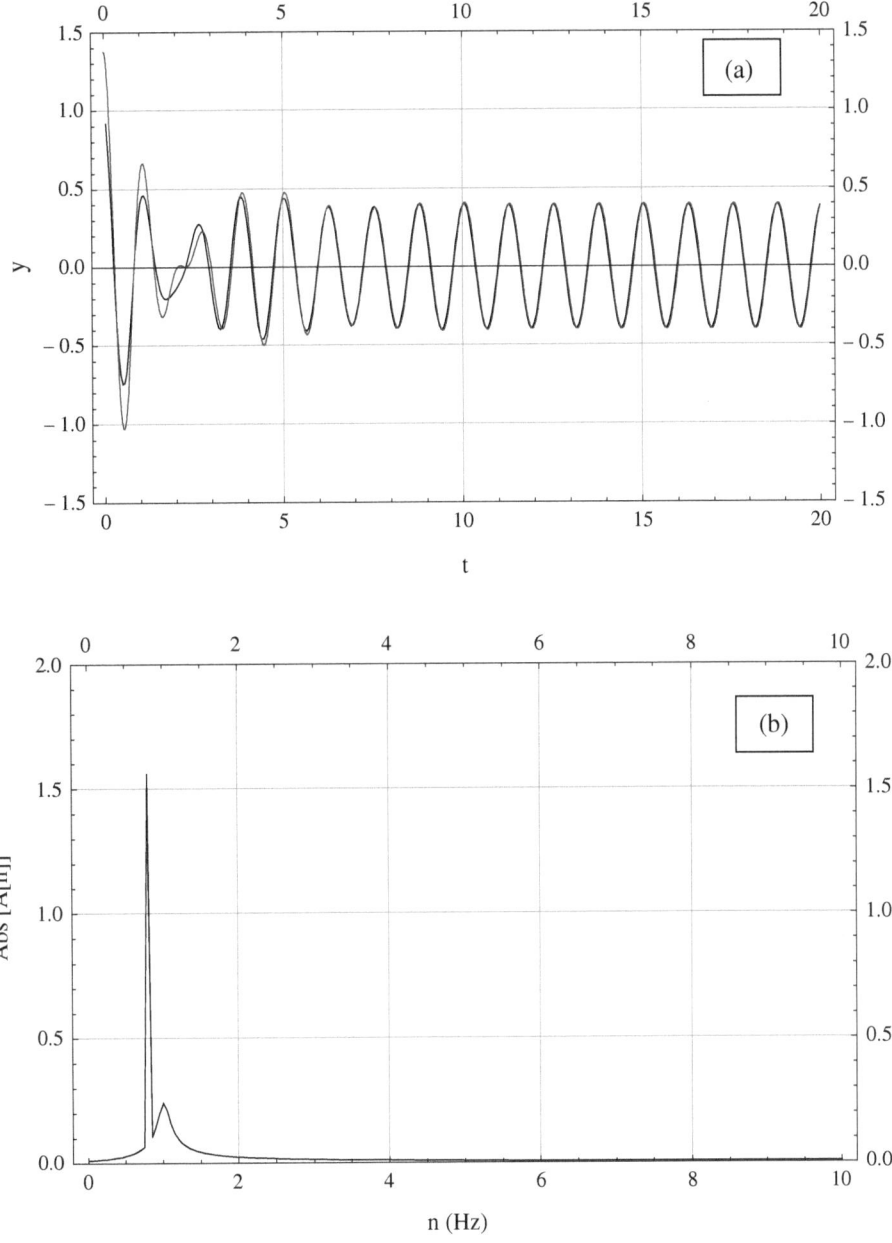

Fig. 2.16 **a** Shows displacement y of driven damped oscillator as a function of time t. Both analytical and numerical results are shown. **b** Shows frequency content in the data of driven damped oscillator, called power spectrum. The parameters are $\omega_0 = 2\pi$; $\omega = 0.8\omega_0$; $\nu = \omega/(2\pi)$, **a = 6**; b $= 1$; $\omega_d = \sqrt{(\omega_0^2 - (b/2)^2)}$; $\nu_d = \omega_d/(2\pi)$, $\nu_d = 0.997$, $\nu = 0.8$

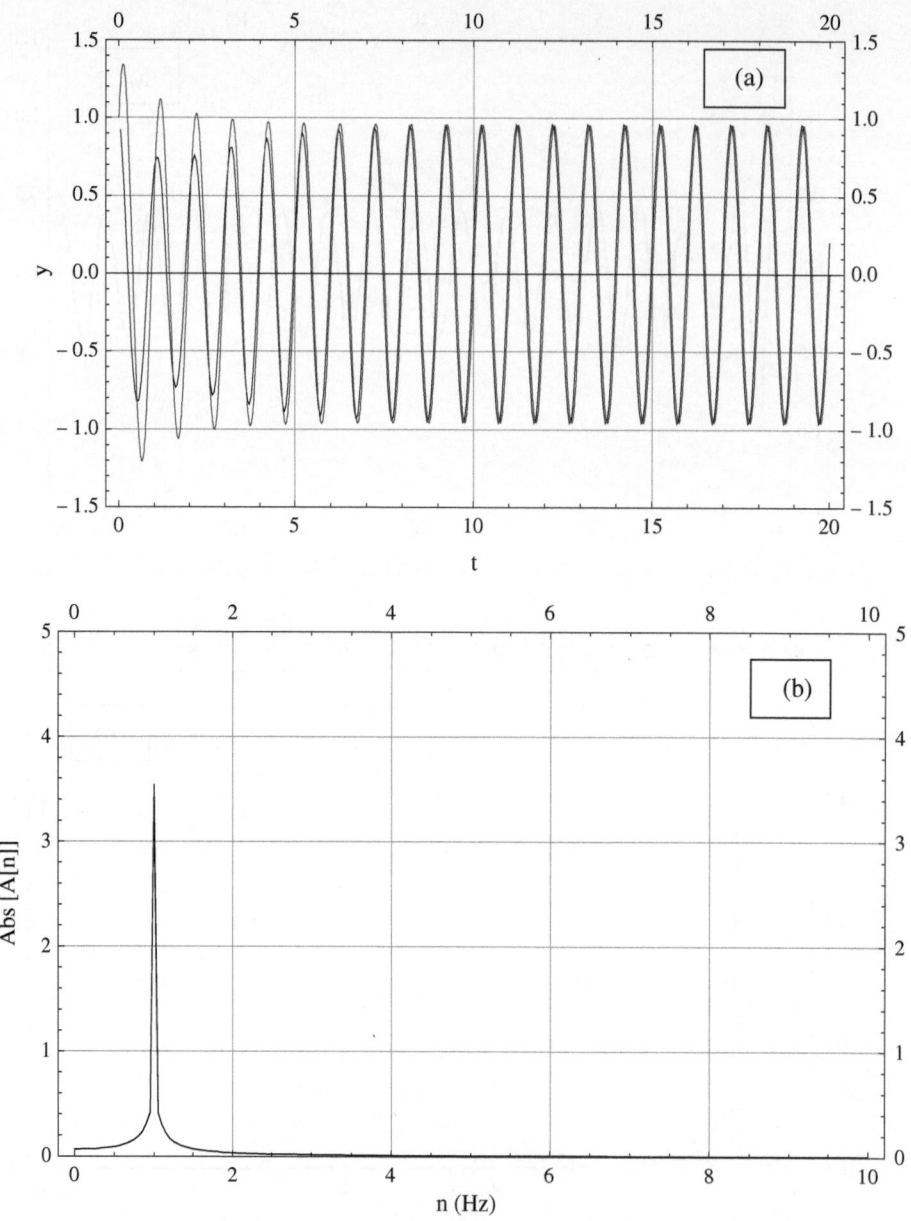

Fig. 2.17 a Shows displacement y of driven damped oscillator as a function of time t. Both analytical and numerical results are shown. **b** Shows frequency content in the data of driven damped oscillator, called power spectrum. The parameters are $\omega_0 = 2\pi$; $\omega = 1.0\omega_0$; $\nu = \omega/(2\pi)$, **a = 6**; b $= 1$; $\omega_d = \sqrt{(\omega_0^2 - (b/2)^2)}$; $\nu_d = \omega_d/(2\pi)$, $\nu_d = 0.997$, $\nu = 1.0$

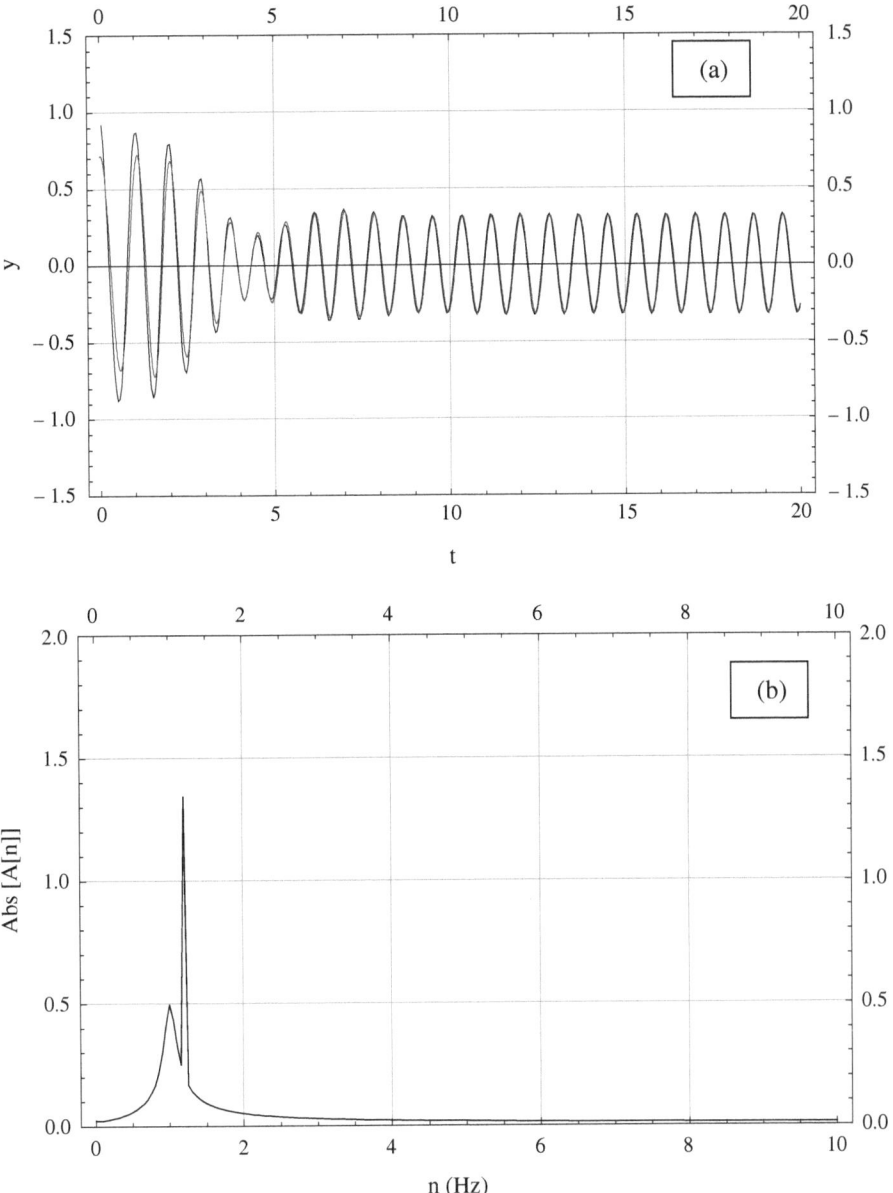

Fig. 2.18 a Shows displacement y of driven damped oscillator as a function of time t. Both analytical and numerical results are shown. **b** Shows frequency content in the data of driven damped oscillator, called power spectrum. The parameters are $\omega_0 = 2\pi$; $\omega = 1.2\omega_0$; $v = \omega/(2\pi)$, **a = 6**; b $= 1$; $\omega_d = \sqrt{(\omega_0^2 - (b/2)^2)}$; $v_d = \omega_d/(2\pi)$, $v_d = 0.997$, $v = 1.2$

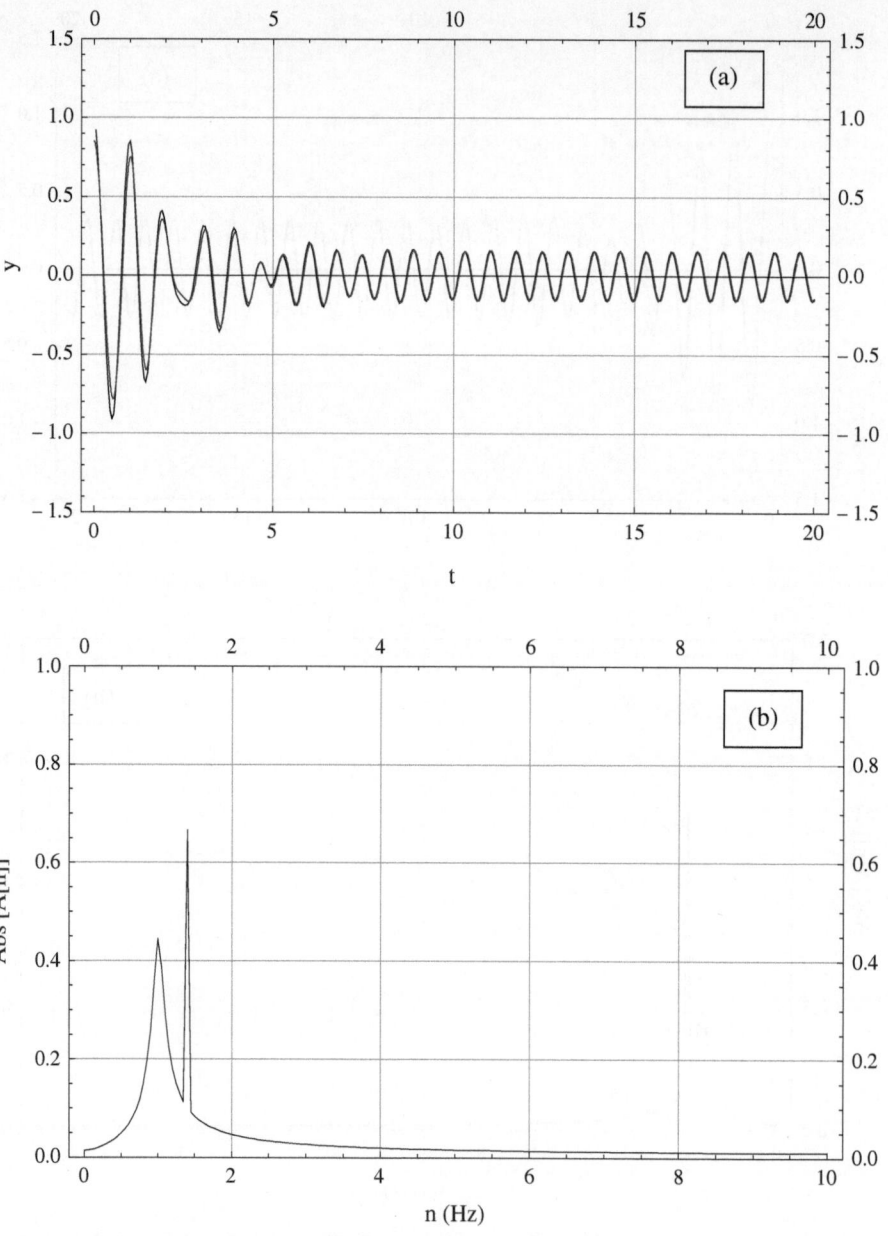

Fig. 2.19 a Shows displacement y of driven damped oscillator as a function of time t. Both analytical and numerical results are shown. **b** Shows frequency content in the data of driven damped oscillator, called power spectrum. The parameters are $\omega_0 = 2\pi$; $\omega = 1.4\omega_0$; $\nu = \omega/(2\pi)$, **a = 6**; b $= 1$; $\omega_d = \sqrt{(\omega_0^2 - (b/2)^2)}$; $\nu_d = \omega_d/(2\pi)$, $\nu_d = 0.997$, $\nu = 1.4$

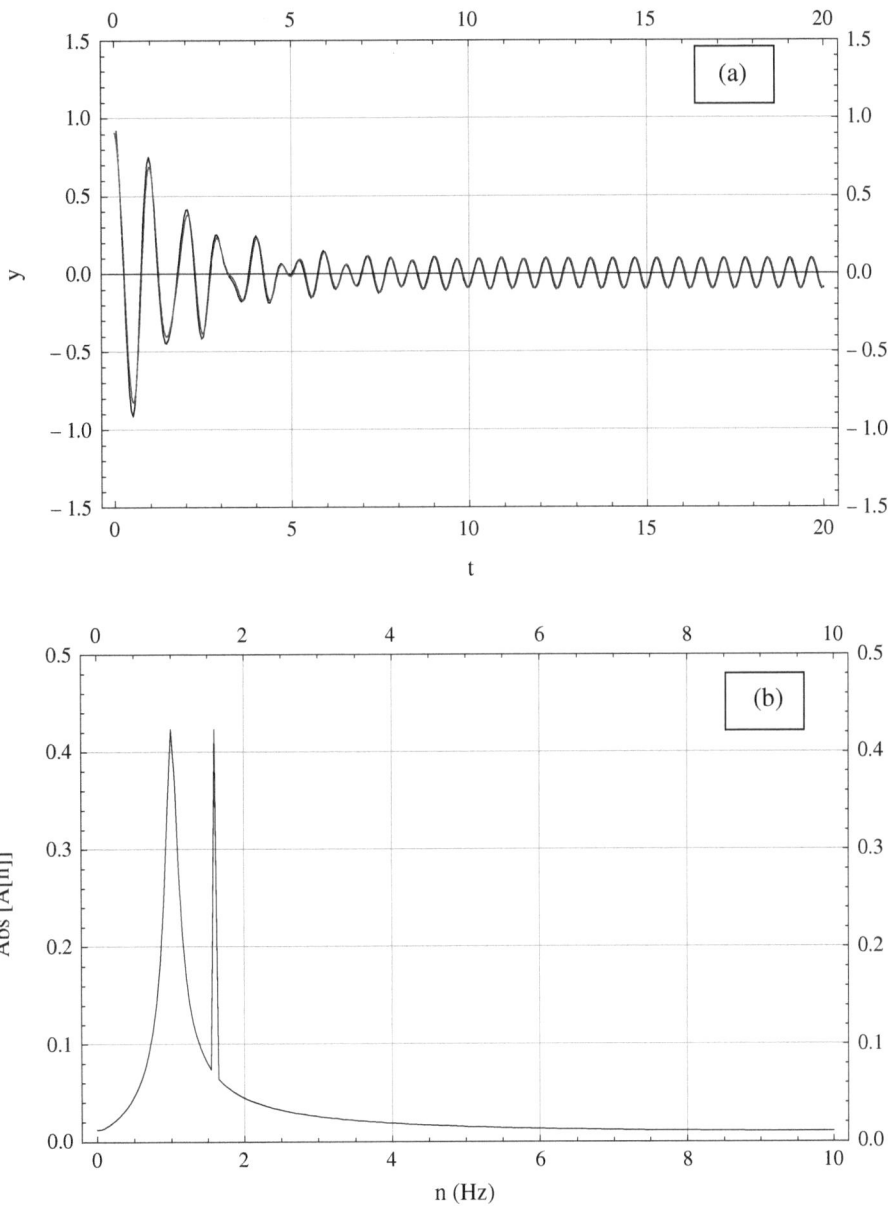

Fig. 2.20 a Shows displacement y of driven damped oscillator as a function of time t. Both analytical and numerical results are shown. **b** Shows frequency content in the data of driven damped oscillator, called power spectrum. The parameters are $\omega_0 = 2\pi$; $\omega = 1.6\omega_0$; $\nu = \omega/(2\pi)$, **a = 6**; b $= 1$; $\omega_d = \sqrt{(\omega_0^2 - (b/2)^2)}$; $\nu_d = \omega_d/(2\pi)$, $\nu_d = 0.997$, $\nu = 1.6$

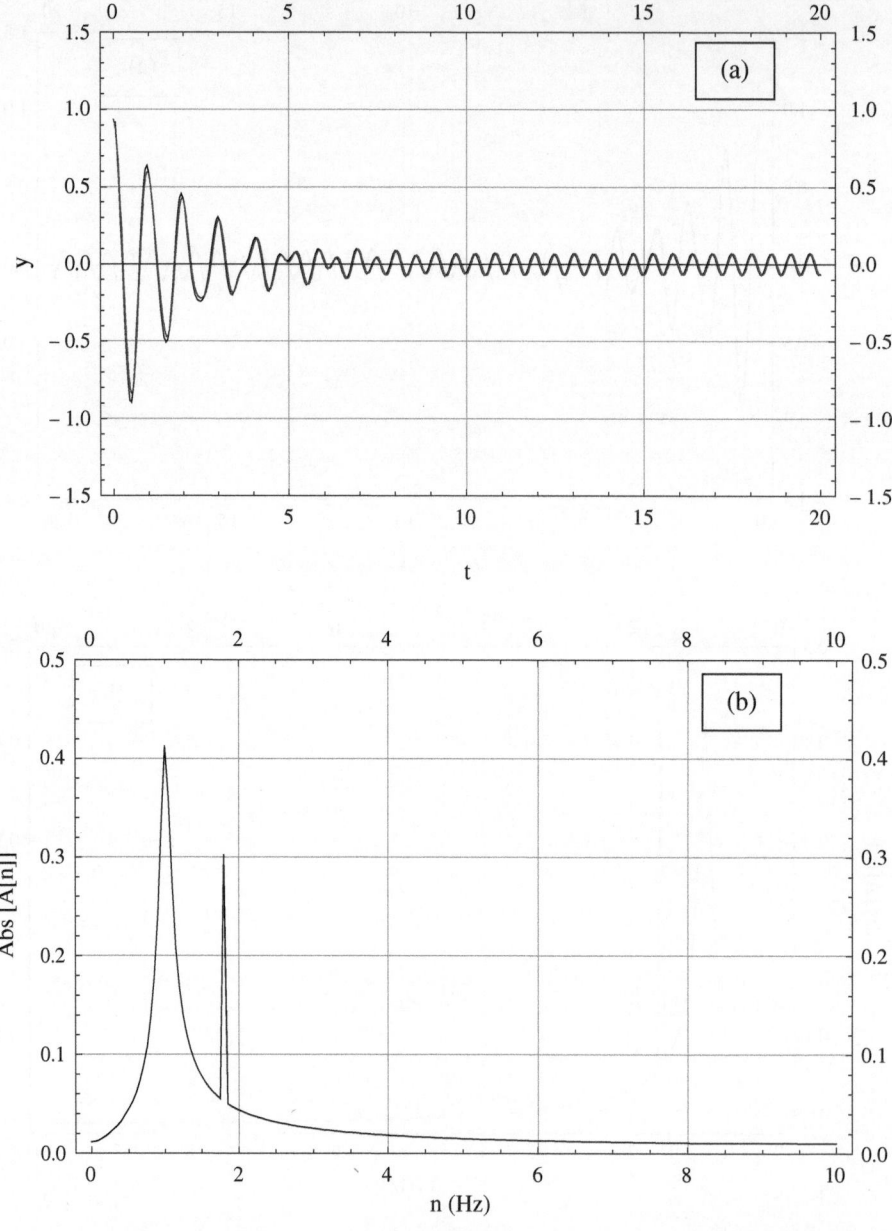

Fig. 2.21 **a** Shows displacement y of driven damped oscillator as a function of time t. Both analytical and numerical results are shown. **b** Shows frequency content in the data of driven damped oscillator, called power spectrum. The parameters are $\omega_0 = 2\pi$; $\omega = 1.8\omega_0$; $\nu = \omega/(2\pi)$, **a = 6**; b $= 1$; $\omega_d = \sqrt{(\omega_0^2 - (b/2)^2)}$; $\nu_d = \omega_d/(2\pi)$, $\nu_d = 0.997$, $\nu = 1.8$

Table 2.4 Amplitude of steady state driven oscillations as a function of $r = \omega/\omega_0$ showing resonance for $r = 1$. For the parameters $\omega_0 = 2\pi$; a = 6; b = 1 used in obtaining Figs. 2.13, 2.14, 2.15, 2.16, 2.17, 2.18, 2.19, 2.20 and 2.21

$r = \omega/\omega_0$	Amplitude
0.2	0.157
0.4	0.182
0.6	0.240
0.8	0.397
1.0	0.957
1.2	0.319
1.4	0.162
1.6	0.103
1.8	0.073

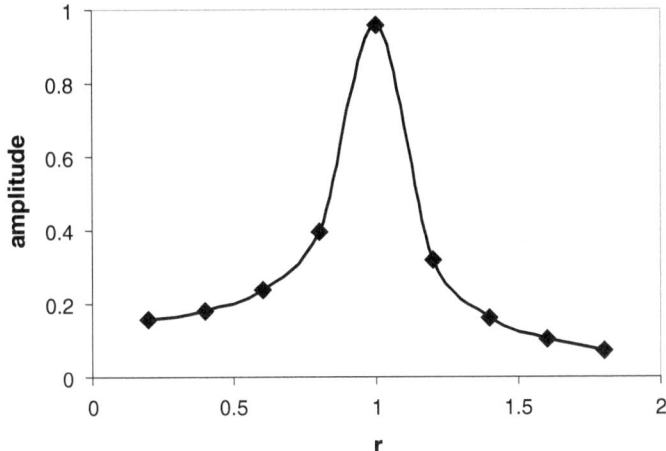

Fig. 2.22 Showing *resonance*. Amplitudes of steady state driven oscillations as function of $r = \omega/\omega_0$. For the parameters $\omega_0 = 2\pi$; **a = 6**; b = 1 used in obtaining Figs. 2.13, 2.14, 2.15, 2.16, 2.17, 2.18, 2.19, 2.20 and 2.21. Using Control+D command in Mathematica, we found a cursor using which we obtained the amplitudes from Figs. 2.13, 2.14, 2.15, 2.16, 2.17, 2.18, 2.19, 2.20 and 2.21 graphically. Because of the larger value of $a = 6$ rather than 3, the amplitude at resonance is larger, ≈ 1 rather than ≈ 0.5 in Fig. 2.12

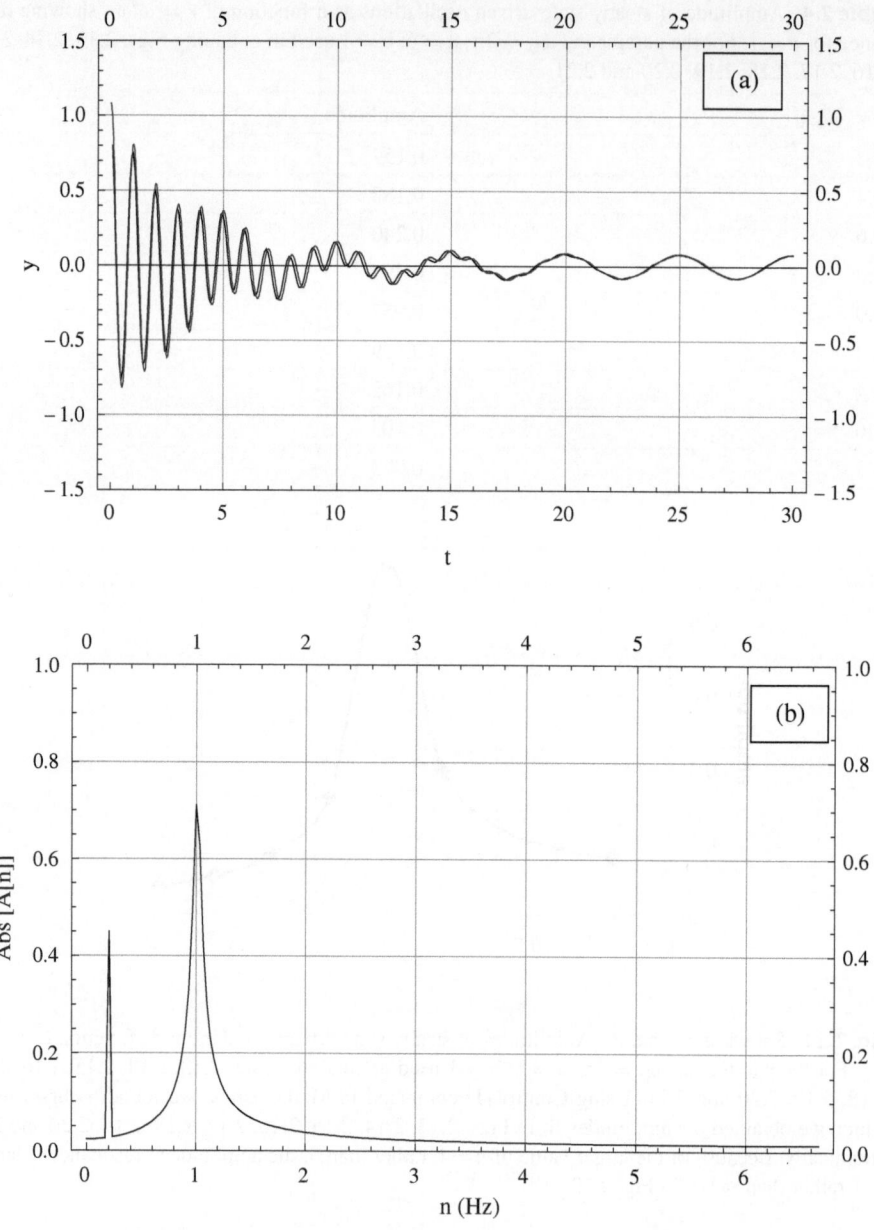

Fig. 2.23 **a** Shows displacement y of driven damped oscillator as a function of time t. Both analytical and numerical results are shown. **b** Shows frequency content in the data of driven damped oscillator, called power spectrum. The parameters are $\omega_0 = 2\pi$; $\omega = 0.2\omega_0$; $\nu = \omega/(2\pi)$, a = 3; **b** = 0.5; $\omega_d = \sqrt{(\omega_0^2 - (b/2)^2)}$; $\nu_d = \omega_d/(2\pi)$, $\nu_d = 0.999$, $\nu = 0.2$

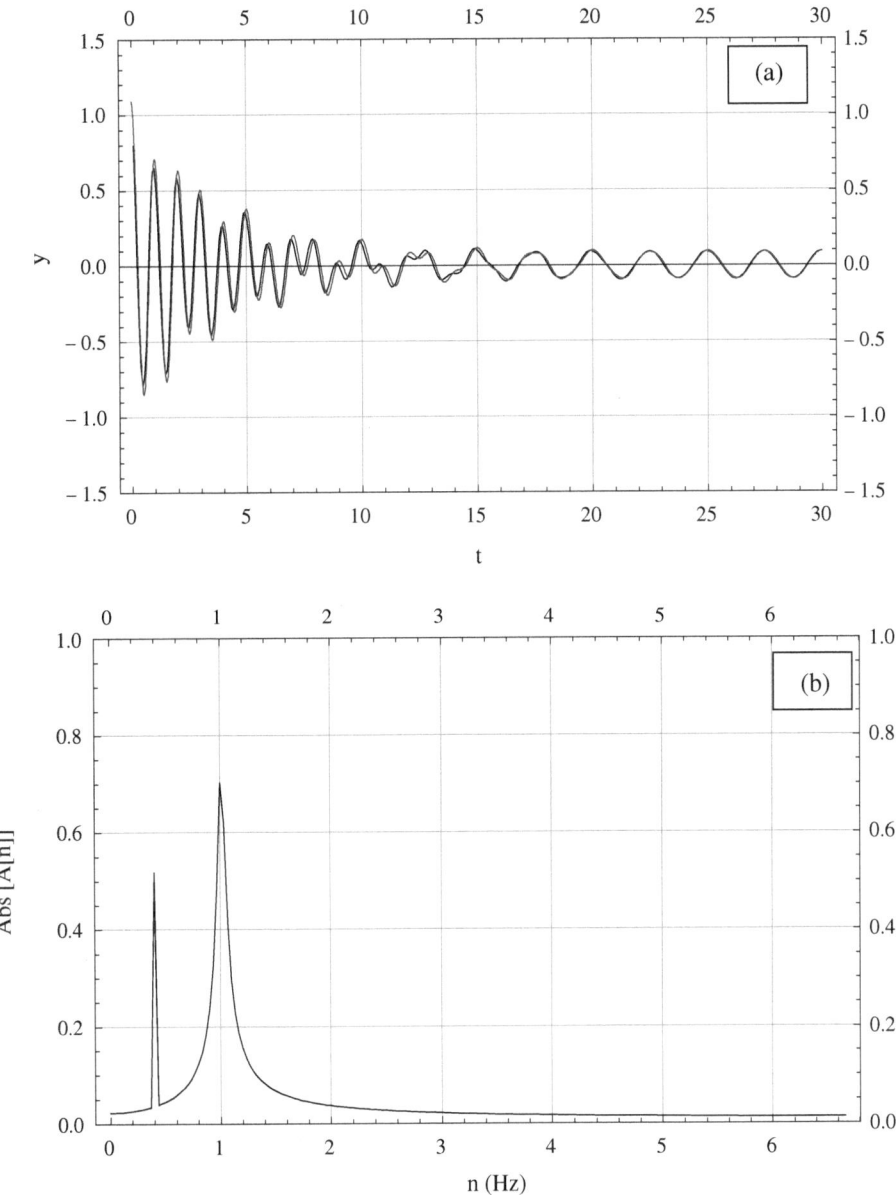

Fig. 2.24 **a** Shows displacement y of driven damped oscillator as a function of time t. Both analytical and numerical results are shown. **b** Shows frequency content in the data of driven damped oscillator, called power spectrum. The parameters are $\omega_0 = 2\pi$; $\omega = 0.4\omega_0$; $v = \omega/(2\pi)$, a $= 3$; **b** $= 0.5$; $\omega_d = \sqrt{(\omega_0^2 - (b/2)^2)}$; $v_d = \omega_d/(2\pi)$, $v_d = 0.999$, $v = 0.4$

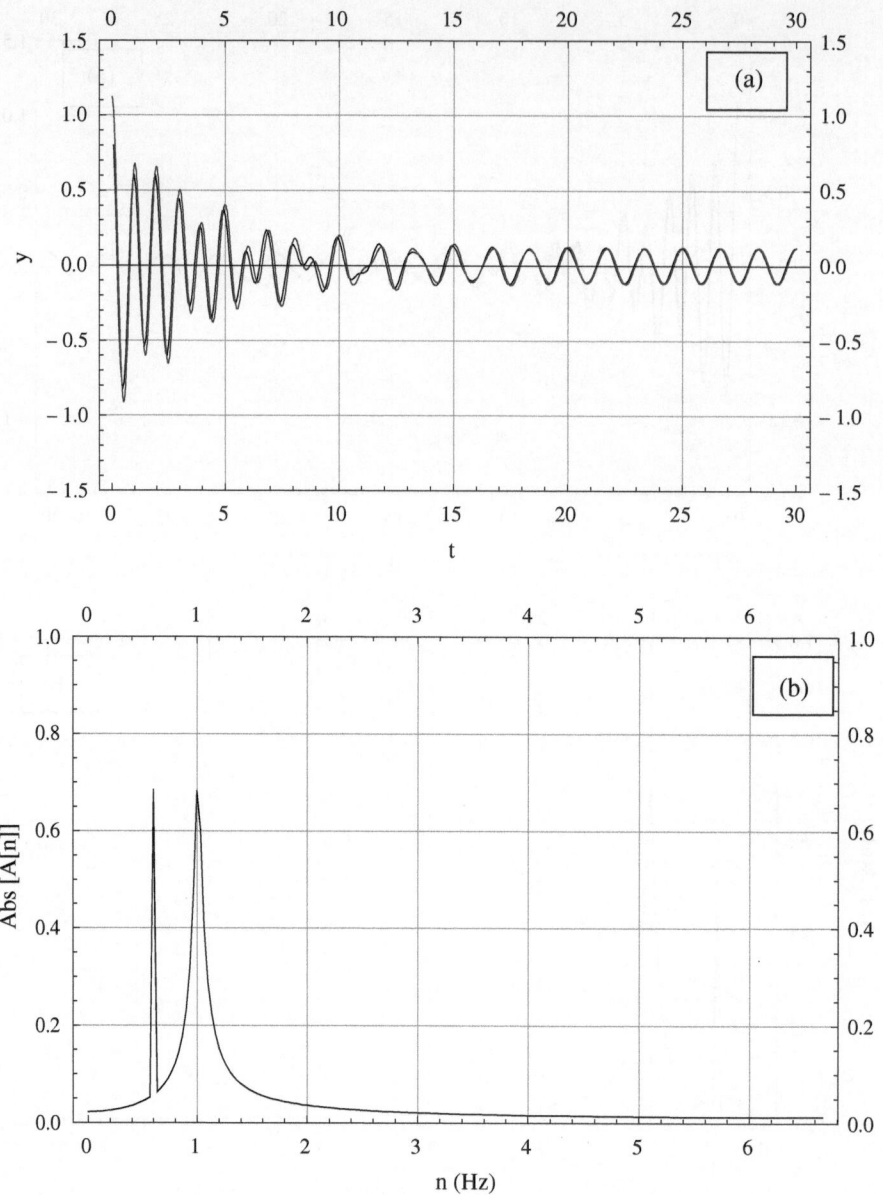

Fig. 2.25 **a** Shows displacement y of driven damped oscillator as a function of time t. Both analytical and numerical results are shown. **b** Shows frequency content in the data of driven damped oscillator, called power spectrum. The parameters are $\omega_0 = 2\pi$; $\omega = 0.6\omega_0$; $\nu = \omega/(2\pi)$, a = 3; **b = 0.5**; $\omega_d = \sqrt{(\omega_0^2 - (b/2)^2)}$; $\nu_d = \omega_d/(2\pi)$, $\nu_d = 0.999$, $\nu = 0.6$

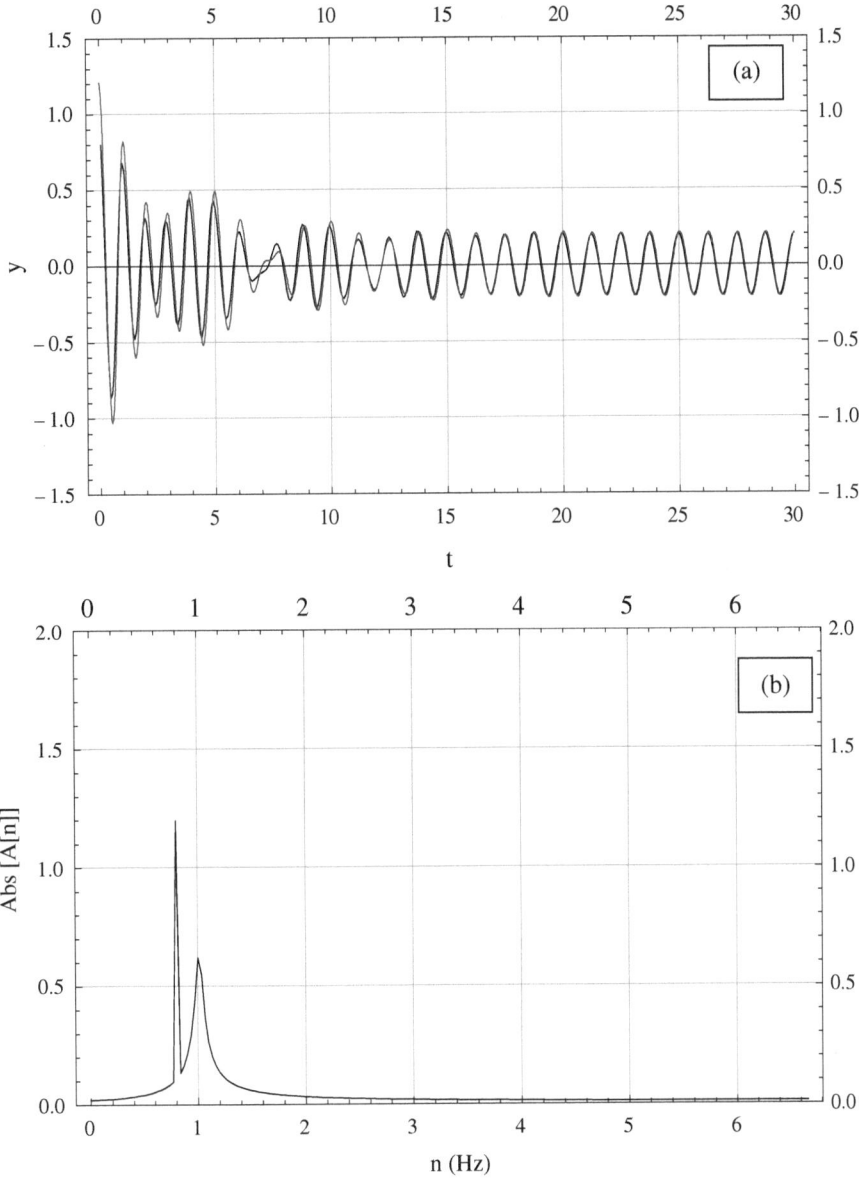

Fig. 2.26 **a** Shows displacement y of driven damped oscillator as a function of time t. Both analytical and numerical results are shown. **b** Shows frequency content in the data of driven damped oscillator, called power spectrum. The parameters are $\omega_0 = 2\pi$; $\omega = 0.8\omega_0$; $\nu = \omega/(2\pi)$, $a = 3$; **b** $= 0.5$; $\omega_d = \sqrt{(\omega_0^2 - (b/2)^2)}$; $\nu_d = \omega_d/(2\pi)$, $\nu_d = 0.999$, $\nu = 0.8$

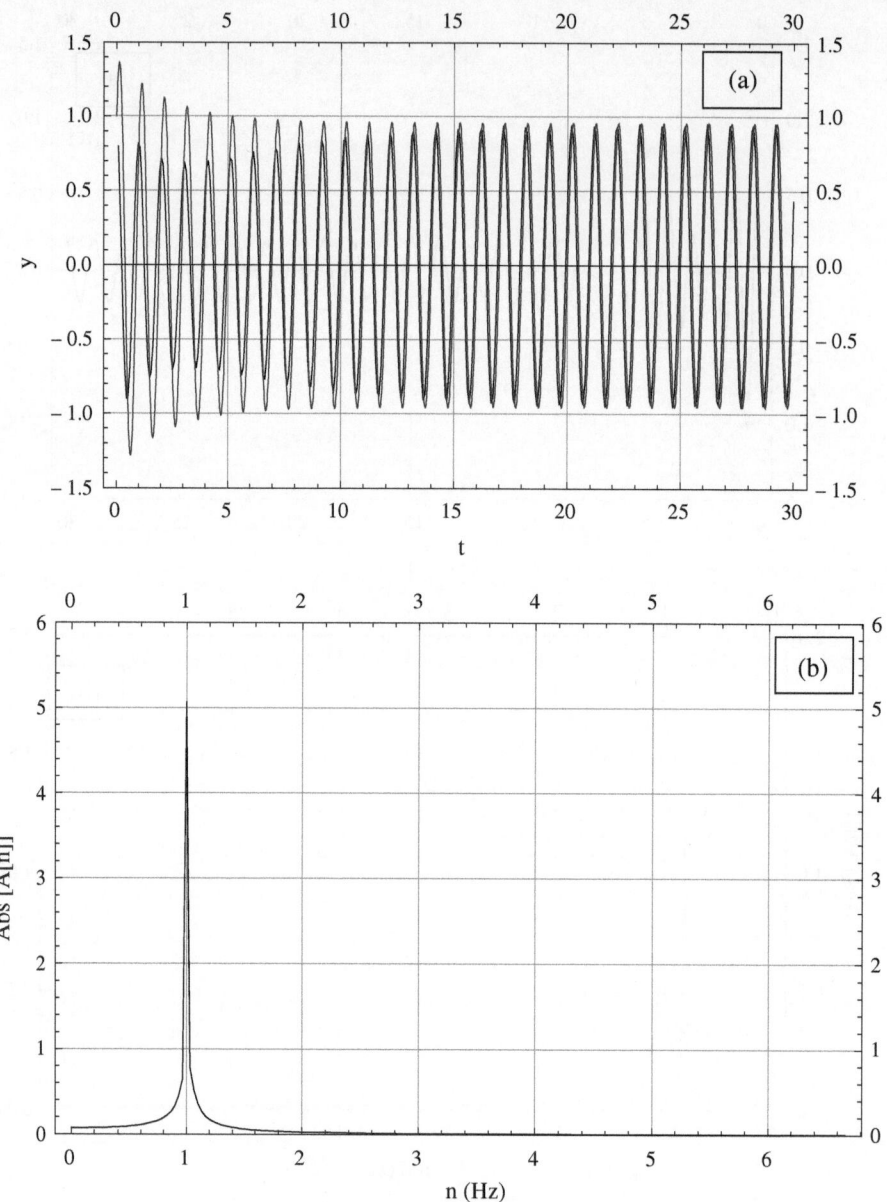

Fig. 2.27 **a** Shows displacement y of driven damped oscillator as a function of time t. Both analytical and numerical results are shown. **b** Shows frequency content in the data of driven damped oscillator, called power spectrum. The parameters are $\omega_0 = 2\pi$; $\omega = 1.0\omega_0$; $\nu = \omega/(2\pi)$, a = 3; **b = 0.5**; $\omega_d = \sqrt{(\omega_0^2 - (b/2)^2)}$; $\nu_d = \omega_d/(2\pi)$, $\nu_d = 0.999$, $\nu = 1.0$

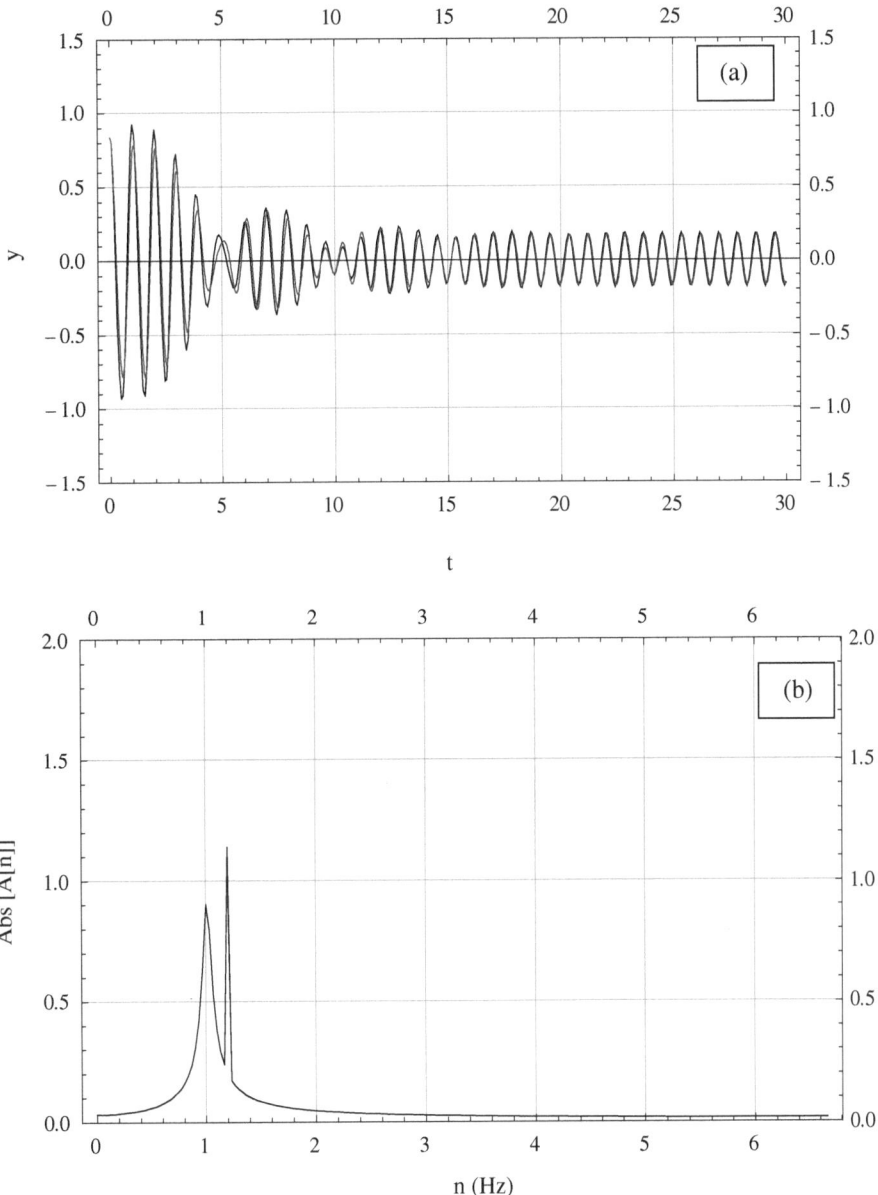

Fig. 2.28 a Shows displacement y of driven damped oscillator as a function of time t. Both analytical and numerical results are shown. **b** Shows frequency content in the data of driven damped
oscillator, called power spectrum. The parameters are $\omega_0 = 2\pi$; $\omega = 1.2\omega_0$; $\nu = \omega/(2\pi)$, a = 3; **b**
$= 0.5$; $\omega_d = \sqrt{(\omega_0^2 - (b/2)^2)}$; $\nu_d = \omega_d/(2\pi)$, $\nu_d = 0.999$, $\nu = 1.2$

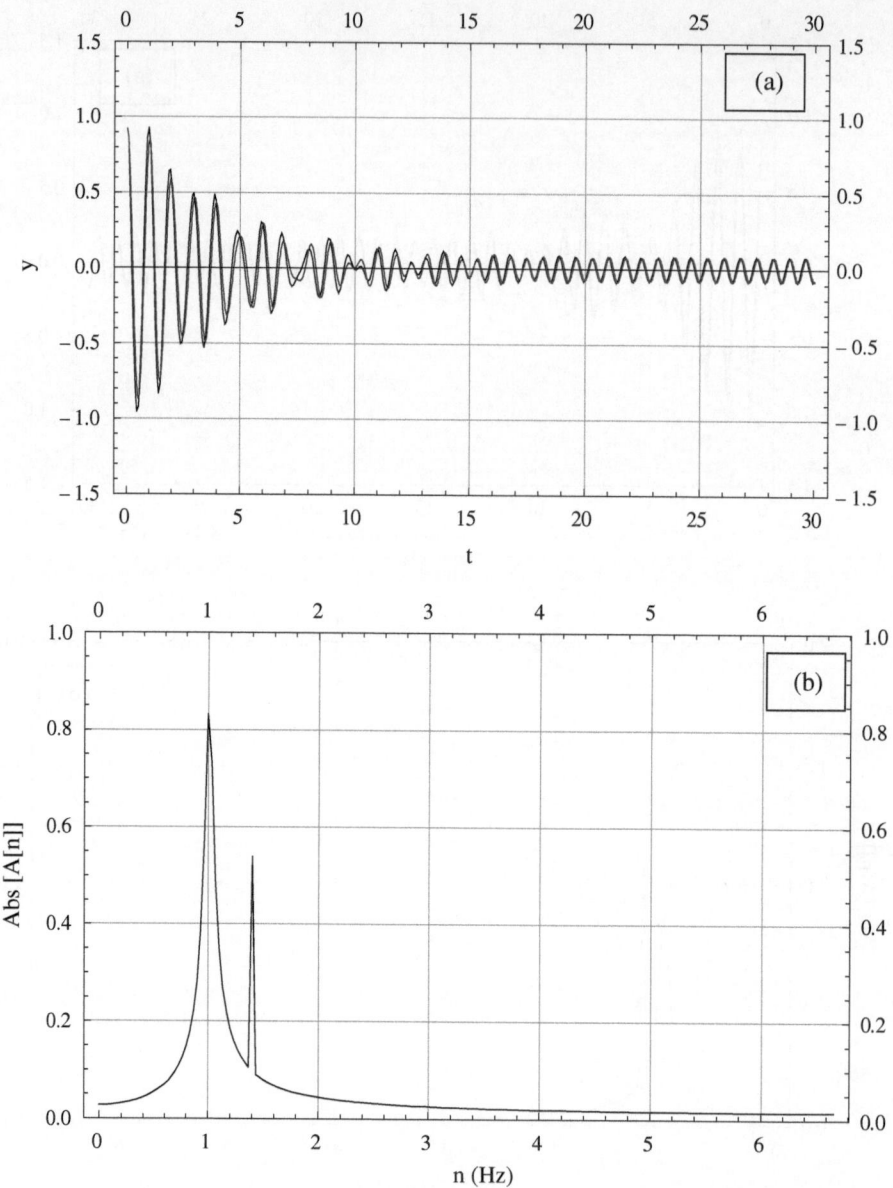

Fig. 2.29 **a** Shows displacement y of driven damped oscillator as a function of time t. Both analytical and numerical results are shown. **b** Shows frequency content in the data of driven damped oscillator, called power spectrum. The parameters are $\omega_0 = 2\pi$; $\omega = 1.4\omega_0$; $\nu = \omega/(2\pi)$, a = 3; **b** = **0.5**; $\omega_d = \sqrt{(\omega_0^2 - (b/2)^2)}$; $\nu_d = \omega_d/(2\pi)$, $\nu_d = 0.999$, $\nu = 1.4$

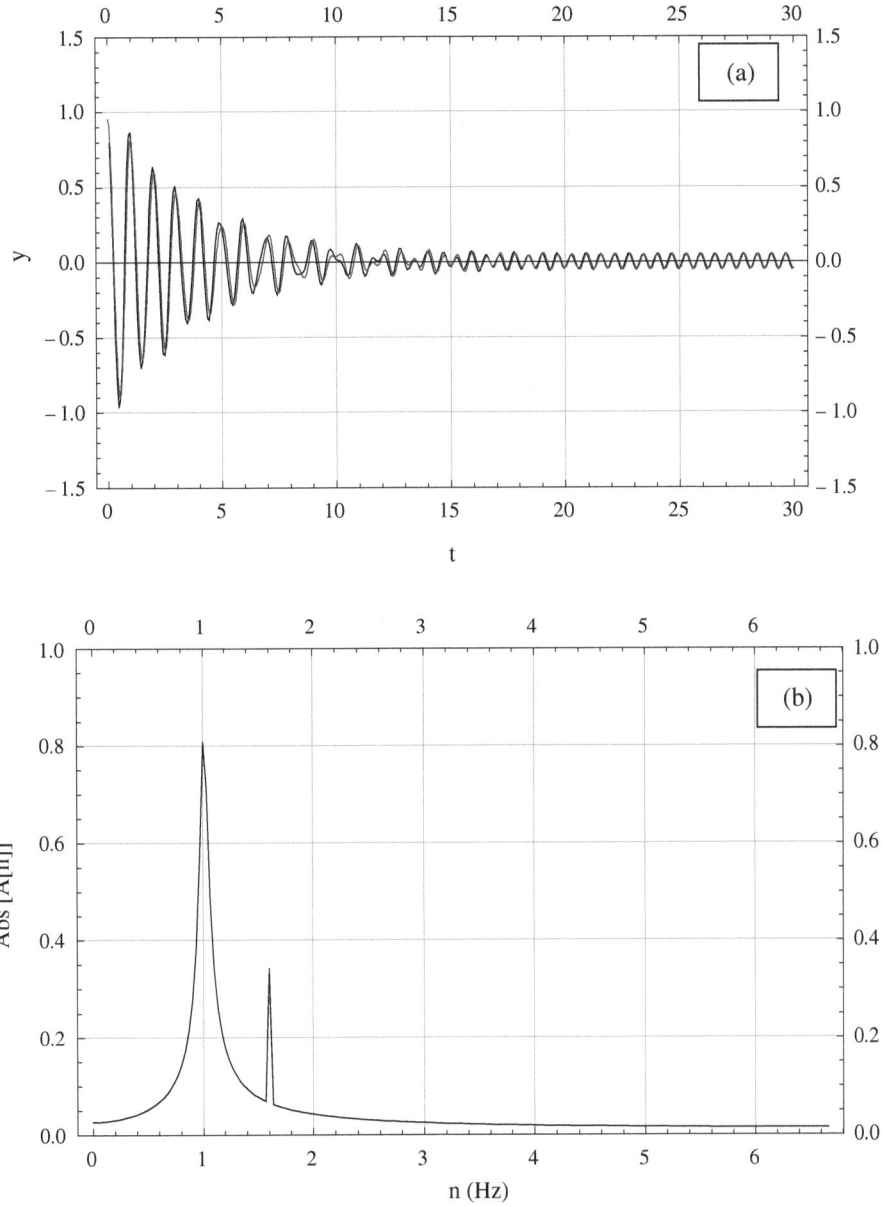

Fig. 2.30 a Shows displacement y of driven damped oscillator as a function of time t. Both analytical and numerical results are shown. **b** Shows frequency content in the data of driven damped
oscillator, called power spectrum. The parameters are $\omega_0 = 2\pi$; $\omega = 1.6\omega_0$; $\nu = \omega/(2\pi)$, a = 3; **b**
= 0.5; $\omega_d = \sqrt{(\omega_0^2 - (b/2)^2)}$; $\nu_d = \omega_d/(2\pi)$, $\nu_d = 0.999$, $\nu = 1.6$

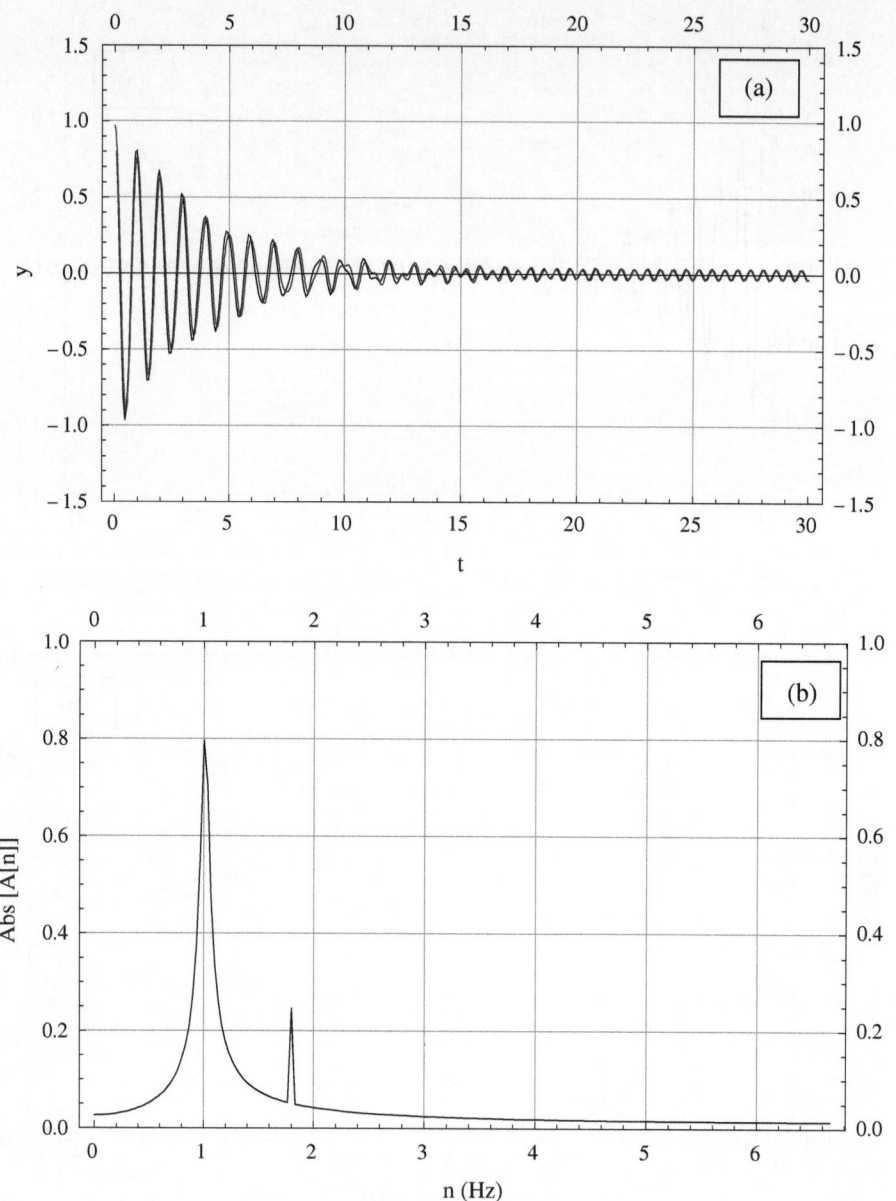

Fig. 2.31 **a** Shows displacement y of driven damped oscillator as a function of time t. Both analytical and numerical results are shown. **b** Shows frequency content in the data of driven damped oscillator, called power spectrum. The parameters are $\omega_0 = 2\pi$; $\omega = 1.8\omega_0$; $\nu = \omega/(2\pi)$, a = 3; **b = 0.5**; $\omega_d = \sqrt{(\omega_0^2 - (b/2)^2)}$; $\nu_d = \omega_d/(2\pi)$, $\nu_d = 0.999$, $\nu = 1.8$

Table 2.5 Amplitude of steady state driven oscillations as a function of $r = \omega/\omega_0$ showing resonance for $r = 1$. For the parameters $\omega_0 = 2\pi$; a = 3; **b = 0.5** used in obtaining Figs. 2.23, 2.24, 2.25, 2.26, 2.27, 2.28, 2.29, 2.30 and 2.31

$r = \omega/\omega_0$	Amplitude
0.2	0.083
0.4	0.092
0.6	0.121
0.8	0.206
1.0	0.951
1.2	0.182
1.4	0.088
1.6	0.055
1.8	0.041

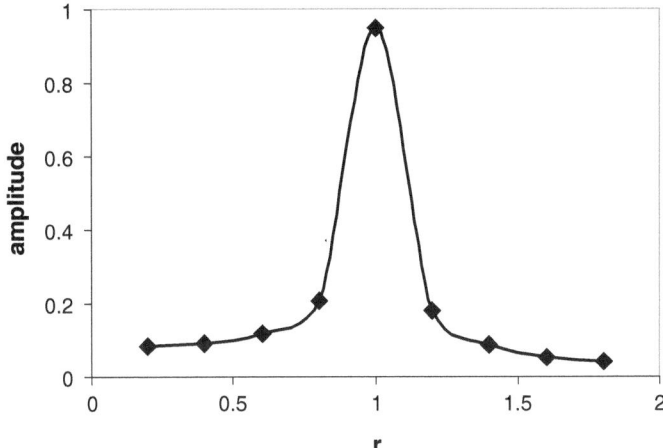

Fig. 2.32 Showing *resonance*. Amplitude of steady state driven oscillations as a function of $r = \omega/\omega_0$. For the parameters $\omega_0 = 2\pi$; a = 3; b = 0.5 used in obtaining Figs. 2.23, 2.24, 2.25, 2.26, 2.27, 2.28, 2.29, 2.30 and 2.31. Using Control+D command in Mathematica, we found a cursor using which we obtained the amplitude from Figs. 2.23, 2.24, 2.25, 2.26, 2.27, 2.28, 2.29, 2.30 and 2.31 graphically. Because of the smaller value of $b = 0.5$ rather than 1, the amplitude at resonance is larger, ≈ 1 rather than ≈ 0.5 in Fig. 2.12

Program number 2.2

```
w0=2*Pi*1.0;
w=0.6*w0;
nu=w/(2*Pi)
a=3;
b=0.5;
wd=Sqrt[w0^2-(b/2)^2];
nud=N[wd/(2*Pi)]


h=N[30/400];
t=0;
v=0;
y=1;
i=0;
Table[{i=i+1,t=t+h,
k1v=h*((-w0^2)*y-b*v+a*Cos[w*(t-h)]);
k2v=h*((-w0^2)*y-b*(v+k1v/2)+a*Cos[w*(t-h+h/2)]);
k3v=h*((-w0^2)*y-b*(v+k2v/2)+a*Cos[w*(t-h+h/2)]);
k4v=h*((-w0^2)*y-b*(v+k3v)+a*Cos[w*(t-h+h)]);
v=v+(k1v+2*k2v+2*k3v+k4v)/6;
k1y=h*(v);k2y=h*(v);k3y=h*(v);k4y=h*(v);
yd[i]=y=y+(k1y+2*k2y+2*k3y+k4y)/6},{i,0,399,1}];
TableForm[%,TableSpacing->{2,2},TableHeadings->{None,{"i","t","y"}}]


t=0;
v=0;
y=1;
i=0;
p1=ListLinePlot[Table[{i=i+1;t=t+h,yd[i]},{i,0,399,1}],
Frame->True,FrameLabel->{"t","y"},PlotStyle->{Black},
PlotRange->{-1.5,1.5},GridLines->Automatic,FrameTicks->All];
p2=Plot[(Exp[-b*t/2])*Cos[wd*t]+
(a/Sqrt[(w0^2-w^2)^2+(b*w)^2])*Sin[w*t-ArcTan[(w^2-w0^2)/(b*w)]],
{t,0,30},PlotRange->{-1.5,1.5},PlotStyle->{Blue}];
```

Show[p1,p2]

Z=N[Exp[-2*Pi*I/400]];

Table[{n1=n1+1,n=N[n1/30],A[n1]=h*(1/Sqrt[2*Pi])*(Sum[yd[k]*(Z^(n1*k)),
{k,1,400,1}]),Ab[n1]=Abs[A[n1]]},{n1,0-1,400-1,1}];
TableForm[%,TableSpacing->{3,3},
TableHeadings->{None,{"n1","n","A[n]","Abs [A[n]]"}}]

ListLinePlot[Table[{n1=n1+1;n=n1/30,Ab[n1]},{n1,0-1,200-1,1}],
Frame->True,FrameLabel->{"n (Hz)","Abs [A[n]]"},PlotStyle->{Black},
GridLines->Automatic,FrameTicks->All,PlotRange->{0,1.0}]

ListLinePlot[Table[{n1=n1+1;n=n1/30,Ab[n1]},{n1,200-1,400-1,1}],
Frame->True,FrameLabel->{"n (Hz)","Abs [A[n]]"},PlotStyle->{Black},
GridLines->Automatic,FrameTicks->All,PlotRange->{0,1.0}]

In the displacement y versus time t plots, we first get damped oscillations with a given frequency $\omega_d/(2\pi)$. Thereafter, we get oscillations of steady amplitude but of a different frequency $\omega/(2\pi)$, that of the driving force.

In the power spectrum, we find presence of 2 dominant frequencies; one of damped oscillations and another of driving force. We find that relative dominance of the 2 frequencies depends on $r = \omega/\omega_0$.

Motion of Driven Damped Oscillator in Phase Space

Abstract

By numerically solving differential equation of motion using 4th order Runge–Kutta method, in this chapter, we have numerically plotted trajectories of simple harmonic oscillator, damped harmonic oscillator and driven damped harmonic oscillator in phase space.

Behavior of mechanical systems such as driven damped oscillator is also understood by plotting velocity versus displacement $v(y)$, rather than displaying displacement versus time $y(t)$. This v versus y coordinate system is known as phase space. Trajectory in phase space provides another perspective of the system, and often it is more valuable than $y(t)$. By numerically solving differential equation of motion using 4th order Runge–Kutta method, in this chapter, we have numerically plotted trajectories of simple harmonic oscillator, damped harmonic oscillator and driven damped harmonic oscillator in phase space.

3.1   Motion of Driven Damped Oscillator in Phase Space

Figure 3.1 shows trajectory of driven damped oscillator in phase space in absence of both damping and driving obtained using Program number 3.1. We have trajectory for simple harmonic oscillator. We have $a = 0$, $b = 0$. The trajectory is circular and as we find, the system does not gain or lose energy with progress in time.

Figure 3.2 shows trajectory of driven damped oscillator in phase space in presence of damping but in absence of driving, obtained using Program number 3.1. We have

© The Author(s), under exclusive license to Springer Nature Switzerland AG 2024   95
S. Chowdhury and A. Al Sakib, *Numerical Exploration of Fourier Transform
and Fourier Series*, Synthesis Lectures on Mathematics & Statistics,
https://doi.org/10.1007/978-3-031-34664-4_3

Fig. 3.1 Trajectory of driven damped oscillator in phase space in absence of both damping and driving. Parameters are $\omega_0 = 2\pi$; $\omega = 0.6\omega_0$; $\nu = \omega/(2\pi)$, $a = 0$; $b = 0$; $\omega_d = \sqrt{\left(\omega_0^2 - (b/2)^2\right)}$; $\nu_d = \omega_d/(2\pi)$, $\nu_d = 1$, $\nu = 0.6$. Using Program number 3.1 but with $a = 0$ and $b = 0$

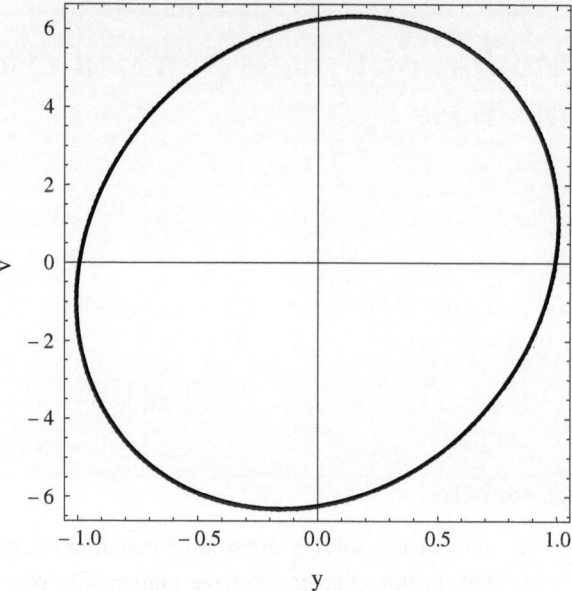

Fig. 3.2 Trajectory of driven damped oscillator in phase space in presence of damping but in absence of driving. Parameters are $\omega_0 = 2\pi$; $\omega = 0.6\omega_0$; $\nu = \omega/(2\pi)$, $a = 0$; $b = 0.5$; $\omega_d = \sqrt{\left(\omega_0^2 - (b/2)^2\right)}$; $\nu_d = \omega_d/(2\pi)$, $\nu_d = 0.9999$, $\nu = 0.6$. Using Program number 3.1 but with $a = 0$ and $b = 0.5$

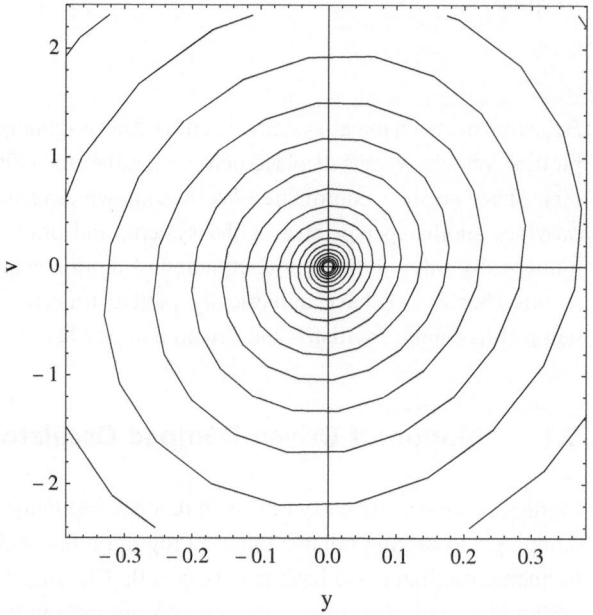

trajectory for damped harmonic oscillator. We have $a = 0$, $b = 0.5$. The trajectory in phase space spirals towards the center as the system loses energy with progress in time.

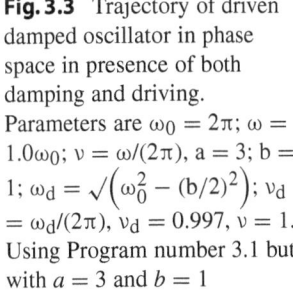

Fig. 3.3 Trajectory of driven damped oscillator in phase space in presence of both damping and driving. Parameters are $\omega_0 = 2\pi$; $\omega = 1.0\omega_0$; $\nu = \omega/(2\pi)$, a = 3; b = 1; $\omega_d = \sqrt{\left(\omega_0^2 - (b/2)^2\right)}$; $\nu_d = \omega_d/(2\pi)$, $\nu_d = 0.997$, $\nu = 1$. Using Program number 3.1 but with $a = 3$ and $b = 1$

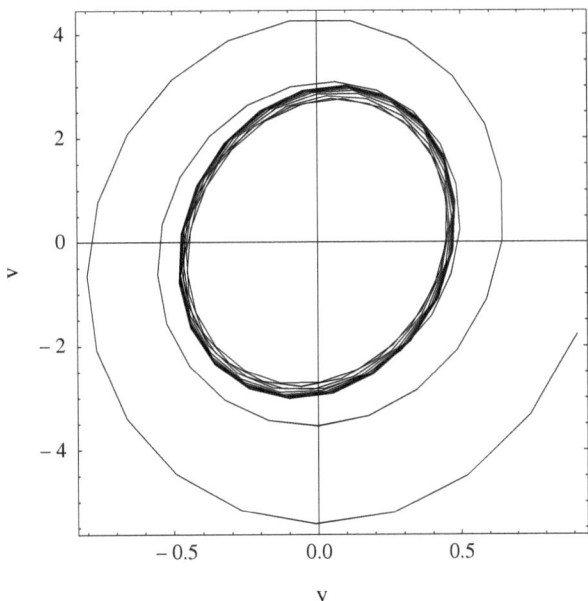

Figure 3.3 shows trajectory of driven damped oscillator in phase space in presence of both damping and driving, obtained using Program number 3.1. We find that trajectory first spirals towards the centre as the system loses energy with progress in time. But soon the driving force takes over and stops the spiral and a steady circle of fixed radius is obtained. We have $a = 3$, $b = 1$.

Program number 3.1

```
w0=2*Pi;
w=0.6*w0;
nu=w/(2*Pi)
a=0;
b=0.5;
wd=Sqrt[w0^2-(b/2)^2];
nud=N[wd/(2*Pi)]


h=N[20/400];
t=0;
v=0;
y=1;
i=0;
```

```
Table[{i=i+1,t=t+h,
k1v=h*((-w0^2)*y-b*v+a*Cos[w*(t-h)]);
k2v=h*((-w0^2)*y-b*(v+k1v/2)+a*Cos[w*(t-h+h/2)]);
k3v=h*((-w0^2)*y-b*(v+k2v/2)+a*Cos[w*(t-h+h/2)]);
k4v=h*((-w0^2)*y-b*(v+k3v)+a*Cos[w*(t-h+h)]);
V[i]=v+ (k1v+2*k2v+2*k3v+k4v)/6;
k1y=h*(v);k2y=h*(v);k3y=h*(v);k4y=h*(v);
yd[i]=y+ (k1y+2*k2y+2*k3y+k4y)/6},{i,0,399,1}];
TableForm[%,TableSpacing->{2,2},
TableHeadings->{None,{ "i","t","y"}}]


t=0;
v=0;
y=1;
i=0;
p1=ListLinePlot[Table[{i=i+1;t=t+h,yd[i]},{i,0,399,1}],
Frame->True,FrameLabel->{"t","y"},PlotStyle->{Black},
PlotRange->{-1.5,1.5},GridLines->Automatic,FrameTicks->All];
p2=Plot[(Exp[-b*t/2])*Cos[wd*t]+
(a/Sqrt[(w0^2-w^2)^2+(b*w)^2])*Sin[w*t-ArcTan[(w^2-w0^2)/(b*w)]],
{t,0,20},PlotRange->{-1.5,1.5},PlotStyle->{Blue}];
Show[p1,p2]


t=0;
v=0;
y=1;
i=0;
p3=ListLinePlot[Table[{i=i+1;yd[i],V[i]},{i,0,399,1}],Frame->True,
PlotRange->Automatic,FrameLabel->{"y","v"},PlotStyle->{Black},
AspectRatio->1]
```

Appendix: Handout for Computational Lab: Determining Frequency Content of Experimental Data

This appendix provides readers handout for computational laboratory. Hands-on demonstration is provided so that readers can obtain frequency content of experimental numerical data. Both Microsoft Excel and Mathematica have been used in the demonstration.

A.1 The Experimental Data

Suppose we have experimental data for y as a function of time t shown in Table A.1. The data can be generated using the function $y = 3/2 + 3 \cos (2\pi\ t) + 4 \sin (6\pi t)$ using Program number A.1. But we now assume that we do not know the function; we just have the numerical data for y as a function of time t shown in Table A.1 obtained by an experiment and we wish to numerically obtain the frequencies that are present in the data; the frequencies are of course 1 and 3 Hz besides 0 corresponding the "dc level" given by 3/2 in the function.

Program number A.1

```
h=N[10/100];
i=0;
Table[{i=i+1,t=t+h,y[i]=3/2+3*Cos[2*Pi*1*t]+4*Sin[2*Pi*3*t]},{t,0,10-h,h}];
TableForm[%,TableSpacing->{3,3},TableHeadings->{None,{"i","t","y[i]"}}]
```

A.2 Gathering the Data

We now use Program number A.2 to generate what is shown in Table A.2 (We need it for Table A.3).

© The Editor(s) (if applicable) and The Author(s), under exclusive license to Springer Nature Switzerland AG 2024
S. Chowdhury and A. Al Sakib, *Numerical Exploration of Fourier Transform and Fourier Series*, Synthesis Lectures on Mathematics & Statistics, https://doi.org/10.1007/978-3-031-34664-4

Table A.1 Showing experimental data for y as a function of time t. We wish to numerically calculate or obtain frequencies that are present in the data

i	t	$y[i]$
1	0.1	7.731280
2	0.2	0.075910
3	0.3	-1.778190
4	0.4	2.877180
5	0.5	-1.500000
6	0.6	-4.731280
7	0.7	2.924090
8	0.8	4.778190
9	0.9	0.122825
10	1.0	4.500000
11	1.1	7.731280
12	1.2	0.075910
13	1.3	-1.778190
14	1.4	2.877180
15	1.5	-1.500000
16	1.6	-4.731280
17	1.7	2.924090
18	1.8	4.778190
19	1.9	0.122825
20	2.0	4.500000
21	2.1	7.731280
22	2.2	0.075910
23	2.3	-1.778190
24	2.4	2.877180
25	2.5	-1.500000
26	2.6	-4.731280
27	2.7	2.924090
28	2.8	4.778190
29	2.9	0.122825
30	3.0	4.500000
31	3.1	7.731280
32	3.2	0.075910
33	3.3	-1.778190

(continued)

Table A.1 (continued)

i	t	y[i]
34	3.4	2.877180
35	3.5	-1.500000
36	3.6	-4.731280
37	3.7	2.924090
38	3.8	4.778190
39	3.9	0.122825
40	4.0	4.500000
41	4.1	7.731280
42	4.2	0.075910
43	4.3	-1.778190
44	4.4	2.877180
45	4.5	-1.500000
46	4.6	-4.731280
47	4.7	2.924090
48	4.8	4.778190
49	4.9	0.122825
50	5.0	4.500000
51	5.1	7.731280
52	5.2	0.075910
53	5.3	-1.778190
54	5.4	2.877180
55	5.5	-1.500000
56	5.6	-4.731280
57	5.7	2.924090
58	5.8	4.778190
59	5.9	0.122825
60	6.0	4.500000
61	6.1	7.731280
62	6.2	0.075910
63	6.3	-1.778190
64	6.4	2.877180
65	6.5	-1.500000
66	6.6	-4.731280

(continued)

Table A.1 (continued)

i	t	y[i]
67	6.7	2.924090
68	6.8	4.778190
69	6.9	0.122825
70	7.0	4.500000
71	7.1	7.731280
72	7.2	0.075910
73	7.3	-1.778190
74	7.4	2.877180
75	7.5	-1.500000
76	7.6	-4.731280
77	7.7	2.924090
78	7.8	4.778190
79	7.9	0.122825
80	8.0	4.500000
81	8.1	7.731280
82	8.2	0.075910
83	8.3	-1.778190
84	8.4	2.877180
85	8.5	-1.500000
86	8.6	-4.731280
87	8.7	2.924090
88	8.8	4.778190
89	8.9	0.122825
90	9.0	4.500000
91	9.1	7.731280
92	9.2	0.075910
93	9.3	-1.778190
94	9.4	2.877180
95	9.5	-1.500000
96	9.6	-4.731280
97	9.7	2.924090
98	9.8	4.778190
99	9.9	0.122825
100	10.0	4.500000

Table A.2 Showing a list of values for y[i] generated using Program number A.2

i	y[i]
1	y[1]
2	y[2]
3	y[3]
4	y[4]
5	y[5]
...	...
96	y[96]
97	y[97]
98	y[98]
99	y[99]
100	y[100]

Table A.3 Showing Microsoft spread-sheet for gathering the values of y. Column B of the spread-sheet can be filled by copying contents of 2nd column of Table A.2. Column D of the spread-sheet can be filled by copying contents of 3rd column of Table A.1. Column C and E of the spread-sheet can be filled by highlighting the cells and using control+D command

	A	B	C	D	E
1	i	y[i]	=	y[i]	;
2	1	y[1]	=	7.731280	;
3	2	y[2]	=	0.075910	;
4	3	y[3]	=	-1.778190	;
...	...	...	...	...	...
101	100	y[100]	=	4.500000	;

Program number A.2

```
i=0;
Table[{i=i+1,y[i]},{i,0,99,1}];
TableForm[%,TableSpacing->{3,3},TableHeadings->{None,{"i","y[i]"}}]
```

We then need to copy contents of the spread-sheet from cell B2 to E101 and paste in Mathematica to get the following:

y[1]=7.731280;y[2]=0.075910;y[3]=-1.778190;y[4]=2.877180;y[5]=-1.500000;
y[6]=-4.731280;y[7]=2.924090;y[8]=4.778190;y[9]=0.122825;y[10]=4.500000;

y[11]=7.731280;y[12]=0.075910;y[13]=-1.778190;y[14]=2.877180;y[15]=-1.500000;
y[16]=-4.731280;y[17]=2.924090;y[18]=4.778190;y[19]=0.122825;y[20]=4.500000;
y[21]=7.731280;y[22]=0.075910;y[23]=-1.778190;y[24]=2.877180;y[25]=-1.500000;
y[26]=-4.731280;y[27]=2.924090;y[28]=4.778190;y[29]=0.122825;y[30]=4.500000;
y[31]=7.731280;y[32]=0.075910;y[33]=-1.778190;y[34]=2.877180;y[35]=-1.500000;
y[36]=-4.731280;y[37]=2.924090;y[38]=4.778190;y[39]=0.122825;y[40]=4.500000;
y[41]=7.731280;y[42]=0.075910;y[43]=-1.778190;y[44]=2.877180;y[45]=-1.500000;
y[46]=-4.731280;y[47]=2.924090;y[48]=4.778190;y[49]=0.122825;y[50]=4.500000;
y[51]=7.731280;y[52]=0.075910;y[53]=-1.778190;y[54]=2.877180;y[55]=-1.500000;
y[56]=-4.731280;y[57]=2.924090;y[58]=4.778190;y[59]=0.122825;y[60]=4.500000;
y[61]=7.731280;y[62]=0.075910;y[63]=-1.778190;y[64]=2.877180;y[65]=-1.500000;
y[66]=-4.731280;y[67]=2.924090;y[68]=4.778190;y[69]=0.122825;y[70]=4.500000;
y[71]=7.731280;y[72]=0.075910;y[73]=-1.778190;y[74]=2.877180;y[75]=-1.500000;
y[76]=-4.731280;y[77]=2.924090;y[78]=4.778190;y[79]=0.122825;y[80]=4.500000;
y[81]=7.731280;y[82]=0.075910;y[83]=-1.778190;y[84]=2.877180;y[85]=-1.500000;
y[86]=-4.731280;y[87]=2.924090;y[88]=4.778190;y[89]=0.122825;y[90]=4.500000;
y[91]=7.731280;y[92]=0.075910;y[93]=-1.778190;y[94]=2.877180;y[95]=-1.500000;
y[96]=-4.731280;y[97]=2.924090;y[98]=4.778190;y[99]=0.122825;y[100]=4.500000;

A.3 Calculating the Frequency Content

We now need to feed the above values to the Fourier transform Program number 1.4 as in Program number A.3. The results are the same as in Sect. 1.5. The frequencies are 0, 1 and 3 Hz.

Program number A.3

h=N[10/100];
y[1]=7.731280;y[2]=0.075910;y[3]=-1.778190;y[4]=2.877180;y[5]=-1.500000;
y[6]=-4.731280;y[7]=2.924090;y[8]=4.778190;y[9]=0.122825;y[10]=4.500000;
y[11]=7.731280;y[12]=0.075910;y[13]=-1.778190;y[14]=2.877180;y[15]=-1.500000;
y[16]=-4.731280;y[17]=2.924090;y[18]=4.778190;y[19]=0.122825;y[20]=4.500000;
y[21]=7.731280;y[22]=0.075910;y[23]=-1.778190;y[24]=2.877180;y[25]=-1.500000;
y[26]=-4.731280;y[27]=2.924090;y[28]=4.778190;y[29]=0.122825;y[30]=4.500000;
y[31]=7.731280;y[32]=0.075910;y[33]=-1.778190;y[34]=2.877180;y[35]=-1.500000;
y[36]=-4.731280;y[37]=2.924090;y[38]=4.778190;y[39]=0.122825;y[40]=4.500000;
y[41]=7.731280;y[42]=0.075910;y[43]=-1.778190;y[44]=2.877180;y[45]=-1.500000;
y[46]=-4.731280;y[47]=2.924090;y[48]=4.778190;y[49]=0.122825;y[50]=4.500000;
y[51]=7.731280;y[52]=0.075910;y[53]=-1.778190;y[54]=2.877180;y[55]=-1.500000;

y[56]=-4.731280;y[57]=2.924090;y[58]=4.778190;y[59]=0.122825;y[60]=4.500000;
y[61]=7.731280;y[62]=0.075910;y[63]=-1.778190;y[64]=2.877180;y[65]=-1.500000;
y[66]=-4.731280;y[67]=2.924090;y[68]=4.778190;y[69]=0.122825;y[70]=4.500000;
y[71]=7.731280;y[72]=0.075910;y[73]=-1.778190;y[74]=2.877180;y[75]=-1.500000;
y[76]=-4.731280;y[77]=2.924090;y[78]=4.778190;y[79]=0.122825;y[80]=4.500000;
y[81]=7.731280;y[82]=0.075910;y[83]=-1.778190;y[84]=2.877180;y[85]=-1.500000;
y[86]=-4.731280;y[87]=2.924090;y[88]=4.778190;y[89]=0.122825;y[90]=4.500000;
y[91]=7.731280;y[92]=0.075910;y[93]=-1.778190;y[94]=2.877180;y[95]=-1.500000;
y[96]=-4.731280;y[97]=2.924090;y[98]=4.778190;y[99]=0.122825;y[100]=4.500000;

Z=N[Exp[-2*Pi*I/100]];

Table[{n1=n1+1;n=N[n1/10],A[n1]=h*(1/Sqrt[2*Pi])*(Sum[y[k]*(Z^(n1*k)),
{k,1,100,1}]),Ab[n1]=Abs[A[n1]]},{n1,0-1,100-1,1}];
TableForm[%,TableSpacing->{3,3},
TableHeadings->{None,{"n","A[n]","Abs [A[n]]"}}]

ListPlot[Table[{n1=n1+1;n=n1/10,Ab[n1]},{n1,0-1,100-1,1}],
Frame->True,FrameLabel->{"n (Hz)","Abs [A[n]]"},PlotStyle->{Black},
GridLines->Automatic,FrameTicks->All,PlotRange->{0,10}]

References

1. Computational Physics: Problem Solving with Python R. H. Landau, M. J. Paez, C. C. Bordeianu 3rd Edition, Wiley-VCH, (2015)
2. Physics Resnick, Halliday and Krane 4th Edition, Volume 1, Wiley (1994), page number 329–331
3. Computational Physics S. Chowdhury American Academic Press (2021)

© The Editor(s) (if applicable) and The Author(s), under exclusive license
to Springer Nature Switzerland AG 2024
S. Chowdhury and A. Al Sakib, *Numerical Exploration of Fourier Transform and Fourier Series*, Synthesis Lectures on Mathematics & Statistics,
https://doi.org/10.1007/978-3-031-34664-4